D1540263

$0075X

THE
PHILOSOPHY
OF
SCIENCE

An Introduction

STEPHEN TOULMIN

HARPER TORCHBOOKS / The Science Library

HARPER & ROW, PUBLISHERS

NEW YORK

THE PHILOSOPHY OF SCIENCE

Printed in the United States of America

First published in 1953 in
the Hutchinson University Library
by Hutchinson & Co. Ltd., London,
and reprinted by arrangement

First HARPER TORCHBOOK edition published 1960

CONTENTS

	Preface	7
I	Introductory	9
II	Discovery	17
III	Laws of Nature	57
IV	Theories and Maps	105
V	Uniformity and Determinism	140
	Suggested Reading	171
	Index	173

PREFACE

SCIENCE and philosophy meet at innumerable points, and are related in countless ways. The philosophy of science has, accordingly, been taken to cover a wide variety of things, ranging from a branch of symbolic logic to the propagation of secularist gospels. Writing a brief introduction to such an amorphous subject is a task of some delicacy, since, in order to avoid being completely superficial, one is forced to limit one's field of attention, and so to set up landmarks where at present none are to be found. In making my own selection, I have particularly kept in mind the audience for which this series is intended: the topics chosen and the manner of treatment are primarily designed to meet the needs of University students in philosophy, and assume no special knowledge either of mathematics or of natural science. At the same time, I hope that the book will have its interest for the general reader.

The knot of problems on which I have concentrated seems to me to underlie the whole range of topics constituting "the philosophy of science": without some understanding of these issues one can, for instance, neither assess the relevance of mathematical logic to the sciences, nor appreciate the true status of those "religions without revelation" sometimes built upon them.

At any rate, I have tried wherever possible to deal with the problems the layman finds puzzling when he reads about the exact sciences.

I owe a special debt to the late Professor Ludwig Wittgenstein, and to Professor W. H. Watson, whose book *On Understanding Physics* I have found a continual stimulus. Others whose ideas I have adopted from time to time without specific acknowledgement include J. J. C. Smart, D. Taylor and John Wisdom. Professor H. J. Paton and Professor Gilbert Ryle have read the completed book and made valuable suggestions, which I have in most cases adopted. If other friends with whom I have talked over the problems here discussed recognize their own ideas in the text, I hope they will forgive me for borrowing them, and take the credit themselves.

S.E.T.

October 1952

CHAPTER I

INTRODUCTORY

NOT everyone can be an expert physicist, but everybody likes to have a general grasp of physical ideas. The learned journals and treatises which record the progress of the physical sciences are open only to trained readers—the *Proceedings of the Royal Society* are less readable nowadays than they were in the Royal Society's early days, when Pepys, Dryden and Evelyn were Fellows. In consequence, there have grown up two classes of writings, less needed in those days, on which the non-scientific reader has to rely for his understanding of the physical sciences. For the ordinary man, there are works of popular science, in which the theoretical advances in physics are explained in a way designed to avoid technicalities; and for students of philosophy there are, in addition, books and articles on logic, in which the nature and problems of the physical sciences are discussed under the heading 'Induction and Scientific Method'.

There are, however, certain important questions which both these classes of work leave undiscussed; and, as a result, the defenceless reader tends to get from them a distorted picture of the aims, methods and achievements of the physical sciences. These are questions for which the phrase 'the philosophy of science' has come to be used: it is the task of this book to draw attention to them, to show in part at least how they are to be answered, and to indicate the kinds of misconception which have been generated in the past by leaving them unconsidered.

1.1 *Logic and the physical sciences*

Notice first the topics one finds discussed in books of logic. Induction, Causality, whether the results of the sciences are true or only highly probable, the Uniformity of Nature, the accumulation of confirming instances, Mill's Methods and the prob-

9

ability-calculus: such things form the staple of most expositions. But to anyone with practical experience of the physical sciences there is a curious air of unreality about the results. Lucid, erudite and carefully argued they may be; yet somehow they seem to miss the mark. It is not that the things that are said are untrue or fallacious, but rather that they are irrelevant: the questions which are so impeccably discussed have no bearing on physics. Meanwhile the actual methods of argument physical scientists employ are only rarely examined. French writers on the philosophy of science, Poincaré for instance, at any rate recognize that in this field one must not take too much for granted. English and American writers on the subject tend nowadays, by contrast, to set off on their work assuming that we are all familiar with the things that scientists say and do, and can therefore get on to the really interesting philosophical points that follow.

This attitude exposes one to serious dangers. For if one has too simple an idea of what scientific arguments are like, one may regard as serious philosophical problems questions having no application to the practice of physicists at all. If one takes it for granted, for instance, that laws of nature can be classed for logical purposes with generalizations like "Women are bad drivers" and "Ravens are black", one may conclude that all appeal to such laws must rest on some presupposition about the reliability of generalizations. But unless one sees in some detail what the status of laws of nature in practice is, one cannot decide whether this is a proper conclusion or no. In fact, laws of nature will not easily fit into the traditional array of logical categories, and their discussion calls for a more refined logical classification. Similarly, one can continue to write about 'Causation and its Place in Modern Science' indefinitely, if one fails to notice how rarely the word 'cause' appears in the writings of professional scientists. Yet there are good reasons for this rarity, and to ignore them is again to divorce the philosophical discussion of scientific arguments from the reality.

The student of philosophy therefore needs an introductory guide to the types of argument and method scientists in actual practice employ: in particular, he needs to know how far these arguments and methods are like those which logicians have traditionally considered. How far do the problems the logic books discuss have any bearing on the things working scientists do? Do we want to attack these problems in the customary fashion, and attempt to propound some novel solution; or should we rather see the problems themselves as arising from an over-naïve conception of what the sciences are like? How do physicists in fact decide that an explanation is acceptable? What sort of job must an expression perform to qualify for the title of 'law of nature'; and how do laws of nature differ from hypotheses? Is the difference a matter of our degrees of confidence in the two classes of propositions, or is the distinction drawn on other grounds? Again, how does mathematics come to play so large a part in the physical sciences? And as for those new entities scientists talk so much about—genes, electrons, meson fields and so on—how far are they thought of as really existing, and how far as mere explanatory devices? These are all questions about whose answers it is easy to be mistaken, unless one pays sufficient attention to the actual practice of scientists: one aim of what follows will be to present these features of the physical sciences which must be understood before we can settle such questions.

1.2 *Popular physics and the layman*

The difficulties that arise over books on popular science are rather different. Here there is no doubt that authentic science is being discussed; but the terms in which it is presented are not as explanatory as they at first seem. There is a tendency for a writer in this field to tell us only about the models and conceptions employed in a novel theory; instead of first giving us a firm anchor in the facts which the theory explains, and afterwards showing us in what manner the theory fits these facts. The best the layman can then hope for is a misleadingly unbalanced picture of the theory; while, at worst, he is liable to put the book down more mystified than when he began it.

Recall, for instance, the way in which Sir James Jeans and Sir Arthur Eddington set about popularizing the theories

of modern physics. Too often they did what was comparatively inessential, that is, introduced us to the particular
conceptions and models used in the theories, while failing to
do what is essential, namely, explain in detail the function of
these models, theoretical conceptions and the rest. Eddington's
well-known account of 'the two tables' is a case in point: to
be told that there is not only a common-sense, solid table,
but also a scientific one, mostly consisting of empty space,
does not particularly help one to understand the atomic
theory of matter. The whole reason for accepting the atomic
model is that it helps us to explain things we could not explain
before. Cut off from these phenomena, the model can only
mislead, raising unreal and needless fears about what will
happen when we put the tea-tray down. The same also goes,
regrettably, for many of those pretty pictures which captured
our imaginations: the picture of the electrons in an atom as
like bees in a cathedral, the picture of the brain as a telephone
exchange, and the rest. Regrettably, it can be said, because as
literary devices they certainly have a value and, if they were
not left to stand on their own feet, might genuinely help us to
understand. As things are, however, they act like a searchlight
in the darkness, which picks up here a pinnacle, here a chimney,
and there an attic window: the detail it catches is lit up dazzlingly,
but everything around is thrown into even greater obscurity
and we lose all sense of the proportions of the building.

But this is not the worst that happens. At times the attempt
to popularize a physical theory may even end by unpopular-
izing it. Jeans, for instance, relied on finding a happy analogy
which would by itself bring home to his readers the chief
features of the General Theory of Relativity. And how did
he invite them to think of the Universe? As the three-dimensional
surface of a four-dimensional balloon. The poor layman, who
had been brought up to use the word 'surface' for two-
dimensional things alone, now found himself instructed to
visualize what for him was a self-contradiction, so it was no
wonder if he agreed to Jeans' calling the Universe a mysterious
one. This mystification was also unnecessary. There is no reason
why the principles of the Theory of Relativity should not be

explained in terms the ordinary reader can make something of—Einstein himself does this very well. But Jeans' method defeated its own end: by trying to make the subject too easy and to do with a simile what no simile alone can do, he led many readers to conclude that the whole thing was utterly incomprehensible, and so must be put aside as not for them.

This might suggest that Jeans was just careless, but there is more to it than that. For the fact that he picked on a mode of expression which to the outsider is self-contradictory points to something which the layman needs to be told about the language of physical theories. When a theory is developed, all kinds of phrases which in ordinary life are devoid of meaning are given a use, many familiar terms acquire fresh meanings, and a variety of new terms is introduced to serve the purposes of the theory. A scientist, who learns his physics the hard way, gradually becomes accustomed to using the novel technical terms and the everyday-sounding phrases in the way required; but he may only be half-aware of what is happening—as Professor Born remarks, the building of the language of the sciences is not entirely a conscious process. This has its consequences when the scientist comes to explain some new theory to the layman. For then he may unwittingly use in his exposition terms and turns of phrase which can be understood properly only by someone already familiar with the theory. To a man trained in the use of sophisticated kinds of geometry the phrase 'three-dimensional surface' may no longer be a self-contradiction, but for him to use it in talking to a non-mathematician is to invite incomprehension. And what applies to 'three-dimensional surfaces' applies equally to 'invisible light' and the like: when scientific notions are being popularized, it is necessary to explain the point of such phrases, instead of making an unexplained use of them.

To introduce a distinction we shall find important later: the adoption of a new theory involves a *language-shift*, and one can distinguish between an account of the theory in the new terminology—in 'participant's language'—and an account in which the new terminology is not used but described—an account in 'onlooker's language'. 'Suppose', as Wittgenstein once said, 'that a physicist tells you that he has at last discovered how to see what people look like in the dark, which no one had ever before known. Then you should not be surprised. If he goes on to explain to you that he has discovered how to photograph by infra-red rays, then you have a right to be surprised if you feel like it. But then it is a different kind of surprise, not just a mental whirl. Before he reveals to you the discovery of infra-red photography, you should not just gape at him; you should say, "I do not know what you mean".'

An analogy will help to explain how misconceptions may follow if we attempt to popularize the physical sciences in this way. When we tell children stories at bed-time, we talk to them about all kinds of people—by which is meant not just rich and poor, white and black, beggars and kings, but logically different kinds of people. Some nights we tell them stories from history, other nights ancient myths; sometimes legends, sometimes fables, sometimes accounts of things that we ourselves have done, sometimes stories by contemporary authors. So in bed-time stories Julius Caesar, Hercules, Achilles, the Boy who cried "Wolf!", Uncle George and Winnie-the-Pooh all appear, at first sight, on the same footing. A clever child, no doubt, soon learns to spot from internal evidence what kind of story tonight's story is; and what sort of people its characters are—fabulous, legendary, or historical. But to begin with we have to explain, in asides, what the logical status of each character and story is, saying, "No, there aren't really any talking bears: this is just a made-up story", or "Yes, this really did happen, when my father's father was a boy." Unless the child is told these things in addition to the stories themselves, he may not know how to take them; and thus he may get quite false ideas about the world into which he has been born, about its history, its inhabitants, and the kinds of thing he might encounter one day as he turned the corner of the street. If entertainment alone were needed, the story alone might do. But the risks of misunderstanding are serious, and for real understanding more is needed.

So also in popular science: the layman is not just ignorant of the theories of science, but also unequipped to understand the terms in which a scientist will naturally begin to explain them. To explain the sciences to him by giving him only potted theories and vivid analogies, without a good number of logical asides, is accordingly like telling a child all the sorts of stories we do tell children and not warning him how very different they are: he will not know how much weight to put on the various things that are said, which of the statements about physics are to be taken at their face value, and which of the characters in the stories he could ever hope to meet.

Perhaps the nub of the difficulty is this, that the popularizer has a double aim. For the layman wants to be told about the theories of the sciences in language he can understand; and he also wants to be told about them briefly, 'in a nutshell'. These two demands are bound in practice to conflict. For a major virtue of the language of the sciences is its conciseness. It is always *possible* to say what a scientific theory amounts to without using the technical terms which scientists introduce to serve the purposes of the theory, but one can do so only by talking at very much greater length. If the popularizer is to explain a theory in everyday terms, and at the same time put it in a nutshell, something must be sacrificed: usually the logical asides are the first things to go, and drastic cuts follow in the account of the phenomena the theory is employed to explain. Once this has happened, the layman is given no real entrance to the subject; for unless he is told a good deal about the phenomena a theory is introduced to explain, and what is even more important, just how much further on we are when this 'explanation' has been given, he might as well have been left quite in the dark. Even a real key is of little use if we do not know what rooms it will let us into. And there is no point at all in being told that Einstein has discovered the metaphorical Key to the Universe if we are not also told what sort of thing counts as opening a door with this Key.

Something can be done, however, to remedy this state of affairs. With the help of a few elementary examples, it should be possible to show the common reader some of the more important things he needs to know about the logic of the physical sciences. There is no reason why he need rest content with the idea that physics is a conglomeration of self-contradictions, like 'invisible light' and 'three-dimensional surfaces', and mysteries like 'the curvature of space': armed with the right questions, he can penetrate behind this screen of words to the living subject. For the words of scientists are not always what they seem, and may be misleading taken out of their original context. The vital thing to know is, what sorts of questions need to be asked, if one is to get a satisfactory account of a theory; and this, fortunately, is something which can be shown as well with simple as with sophisticated examples. To show, with illustrations, what these questions are is the principal aim of this book; and it will require us, not so much to quote the things that scientists say, as to see what sort of things they do with the words they employ. As Einstein has said, 'If you want to find out anything from the theoretical physicists about the methods they use, I advise you to stick closely to one principle: don't listen to their words, fix your attention on their deeds.'

CHAPTER II

DISCOVERY

IF we are to know what questions to ask about physical theories, we must be clear to begin with what kinds of things count as discoveries in the physical sciences. What is it for something to be 'discovered' in physics? When a physicist announces that it has been discovered that heat is a form of motion, or that light travels in straight lines, or that X-rays and light-waves are varieties of electro-magnetic radiation, what kind of discovery is this? What does such a discovery amount to?

The question can be put in another way: if, in physics, someone claims to have discovered something, what sort of demonstration will justify us in agreeing that, whereas this was not previously known, it can now be regarded as known? Is it like that required when an explorer discovers a new river, or when a botanist discovers a new variety of flower, or when a doctor discovers what is wrong with a patient, or when an engineer discovers how to bridge a hitherto-unbridgeable river, or when a man doing a crossword-puzzle discovers the word that has been eluding him? Or is it like none of these?

2.1 *Physics presents new ways of regarding old phenomena*

This question will best be answered with the help of examples. Let us look first at a discovery so elementary that it may hardly seem nowadays ever to have needed discovering, or to be anything more than a piece of common sense: the discovery that light travels in straight lines. This example, for all its appearance of obviousness, displays many of the features characteristic of discoveries in the exact sciences. Its very commonsensicality is indeed a merit, reminding us how the sciences grow out of our everyday experience of the world, and what people mean who speak of science, epigrammatically, as 'organized common sense'.

To recognize just what was discovered when it was first announced that 'light travels in straight lines', we must think ourselves back to the way things were before the discovery. This is not entirely easy to do, for we tend nowadays to grow up completely familiar with the idea that sunlight, shadows and the like are the effects of light travelling: it is only by an effort that one can throw off the habit, and look at optical phenomena once again with the eyes of those who knew nothing of geometrical optics, to whom this would be a novel, revolutionary suggestion. Yet the effort is worth making. So let us ask, for a start, what would have been the data on which this discovery was based?

There are three sources of material which we can think of as the backing for it: first, our experience of everyday phenomena like those of light and shade; second, the practical skills and techniques which have been developed as a result of this experience; and third, those regularities in optical phenomena which are not stated but taken for granted and enshrined in our everyday language. We know very well, for instance, that the higher the sun rises in the sky, the shorter are the shadows cast by the objects it illuminates; and that, as it moves across the sky, so do the shadows turn with it. Out of this knowledge, and exploiting it, have grown the techniques used in the design of sundials: the sundial-maker in the course of his trade develops a familiarity with optical phenomena which provides a second starting-point for optics. And there is a further range of physical regularities, with which everyone becomes familiar at an early age, but which are rarely stated. It is harder work running uphill than down; the shortest way to get to the opposite corner of a field is to 'follow your nose'; put your hand in the fire and it will burn you—these are things which any child, and many animals too, may be said to know, yet they seem almost tautologous when put into words; for our recognition of them comes before, rather than after, the development of our everyday language. The way we ordinarily use the word 'straight', for instance, takes it for granted that the shortest and the straightest road are both the one you can see straight along; and our manner of using words like 'up' and 'down', 'fire' and 'burns' likewise links together things we commonly find going together.

The question that faces us is the question, what kind of step is taken when we pass from these data to the conclusion that 'light travels in straight lines'. What type of inference is this? Or is the very word 'inference' a misleading name for such a step?

Let us, as a preliminary, try setting this step alongside a couple of inferential steps, which at first sight it resembles. Robinson Crusoe, we are told, found a footprint on the sandy beach of his island, and concluded that a man had been walking there. Again, a naturalist studying the migration of swallows might find, by plotting the observed tracks of a large number of flocks, that they all flew along 'great circles'. In these cases, too, one can speak of discoveries being made, which can be put in the words "A man has been walking along the beach" and "Migrating swallows always travel along great circles". Let us contrast these discoveries with the discovery that "light travels in straight lines": how does the step from our observations on shadows to this discovery compare with Crusoe's step from the footprint to a man walking, and the naturalist's step from the bird-watcher's reports to his generalization about migrating swallows?

Two important differences spring to the eye at once.

(i) To compare first the step from shadows to light and the step from the footprint to a man. One might turn a corner and come face to face with the man who was responsible for the footprint—this, in fact, was what Robinson Crusoe was terrified of doing. But telling from our study of shadows that light travels in straight lines is quite unlike telling from a footprint that a man has been walking on a beach. To hint at the difference, there is nothing in the case which would count as 'coming face to face with' the light which was 'responsible for' the shadows: no single happening could establish the optical theory once and for all, in the way Crusoe's conclusion could be established. For Crusoe reached his novel conclusion by applying a familiar type of inference to fresh data: "Footprint! Footprints mean man. Therefore man." But in geometrical optics it is not the data which are fresh, for we have known about shadows for a very long time. The novelty of the conclusion comes, not from the data, but from the inference: by it we are led to look at familiar phenomena in a new way, not at new phenomena in a familiar way.

The discovery that light travels in straight lines was not, therefore, the discovery that, where previously nothing had been thought to be, in any ordinary sense, travelling, there turned out on closer inspection to be something travelling— namely, light: to interpret the optical statement in this way would be to misunderstand its point. We can call this the 'Man Friday fallacy'.

(ii) Nor is it the discovery that whatever is travelling, in the everyday sense, is doing so in one way rather than another, along great circles rather than parallels of latitude, or straight lines rather than spirals. Often enough, as we soon find out, light does not travel strictly in straight lines, but is diffracted, refracted or scattered; yet, in practice, this in no way affects the point of the principle that light travels in straight lines (the Principle of the Rectilinear Propagation of Light). In this respect the optical discovery is quite unlike the naturalist's discovery about swallows, which was precisely that they always migrate thus, and not otherwise. Rather, the optical discovery is, in part at any rate, the discovery that one can speak at all profitably of something as travelling in these circumstances, and find a use for inferences and questions suggested by this way of talking about optical phenomena—the very idea that one should talk about anything as travelling in such circumstances being the real novelty.

These differences are, however, only pointers towards a larger difference, and this we must now try to state. In Robinson Crusoe's discovery, and in the naturalist's also, the language in which the conclusion is expressed, like that in which the data would be reported, is the familiar language of everyday life: there is no question of giving new senses to any of the words involved, or of using them in a way which is at all out of the

ordinary. Yet in the optical case, both the key words in our
conclusion— 'light' and 'travelling'—are given new uses in
the very statement of the discovery. Before the discovery is
made, the word 'light' means to us such things as lamps—
the 'light' of "Put out the light"; and illuminated areas—the
'light' of "The sunlight on the garden". Until the discovery,
changes in light and shade, as we ordinarily use the words
(i.e. illuminated regions which move as the sun moves), remain
things primitive, unexplained, to be accepted for what they are.
After the discovery, we see them all as the effects of something,
which we also speak of in a new sense as 'light', travelling from
the sun or lamp to the illuminated objects. A crucial part of the
step we are examining is, then, simply this: coming to think
about shadows and light-patches in a new way, and in consequence coming to ask new questions about them, questions
like "Where from?", "Where to?" and "How fast?", which
are intelligible only if one thinks of the phenomena in this
new way.

It is worth emphasizing how far the physicist's way of looking at optical phenomena *is* a new way, and how far by accepting
it we are required to extend the notions of light and travelling.
Until one has been introduced to the fundamental ideas of
geometrical optics, there is no way of understanding what it
means for a physicist to talk of light travelling: he clearly does
not mean 'sending lanterns by rail', nor can he mean 'cloud-
shadows drifting across the grass' for he talks of light travelling
equally whether the patches of light are moving or still. Indeed,
it would be somewhat queer, in the sort of situations with which
the physicist is concerned, to talk in the ordinary sense of the
word of anything 'travelling' at all.

An example will bring this queerness out. Suppose that
we are sitting on a hillside, gazing across the country, and you
ask, "Is anything on the move?": the appropriate answer will
be some such thing as "Clouds and larks overhead, down below
two men on horseback and a wagon of hay, and there in the
distance a railway-train"—and this answer may be, from the
everyday point of view, an exhaustive one. Taking your question
in the sense in which it was asked, I could neither give nor you

accept, as an answer to it, such a reply as "Photons". It is true that I might say, "Light": but, were I to do so, I could only be understood to mean, e.g., patches of sunlight moving across the heather on the far hillside, and this is certainly not what the physicist means when he speaks of 'light travelling'. And if I *were* to answer "Photons", all you could do would be to wonder whether I did so out of plain misunderstanding; or whether, as the expression of a poetical fancy, I was choosing to borrow a term from physics to suggest, with Heraclitus and Walt Whitman, that even when so few things are, literally, on the move, the world still 'teems with flux'. At any rate— and this is all that it is essential to recognize—the introduction of the notion of 'light' as something 'travelling' is not the simple, literal discovery of something moving, like the detection of frogs in a flower-bed or boys in an apple tree: rather it is an extension of the notion of travelling to do a new job in the service of physics.

Not only is it an extended application of the word: it is also rather a thin one. Somehow, indeed, the use of the particular word 'travelling' does not seem to be of central importance. One finds it being used alongside other words which, from a non-scientist's point of view, are quite incompatible with it: light will be spoken of in the same book sometimes as 'travelling', but at other times as 'being propagated'. Yet there is certainly something of central importance about the kind of word whose meaning it is found natural to extend in this way.[1] So, in answer to the question, "What sort of discovery is this?", we can already give something of a hint: the discovery that light travels in straight lines is, in part at least, the discovery that the phenomena from which we started (shadow-casting and the rest), can be regarded as consequences of something (it matters not yet what) travelling, or being propagated, or something of the kind, from the light-source to the surrounding objects, except where it is cut off by intervening bodies of the kind we call 'opaque'.

[1]The sort of word chosen must reflect such familiar facts as this: that by lighting a lamp in one corner of a room one can produce patches of light in another.

2.2 New points of view come with new inferring techniques

The next question to be asked is this: What does it mean to say that these phenomena *can be regarded* in this way? Still more, what could it mean for a physicist to say, as he might do, that they *must* be so regarded? For, as we have seen, to say this is not like saying that a certain kind of depression in the sand must be the effect of a man standing on it. Since there is nothing quite like meeting Man Friday which would oblige us to accept the new optical theory, how is it that we *must* do so? May we not decline to look at the phenomena in this new way?

Certainly we can. We are not compelled unconditionally to think of the phenomena in the physicist's way. We can, if we choose, refrain from asking any scientific questions about them. If we prefer, we can think of the phenomena of sight as the Greeks did, regarding the eye not as a kind of sensitive plate, but as the source of antennae or tentacles which stretch out and seize on the properties of the objects it surveys. Not only *can* we look at it in this way; we quite frequently do so, or talk as though we did—as, for instance, when we speak of Able Seaman Jones, seated in the crow's-nest, 'sweeping the horizon' with his eagle eyes. Outside physics, the way we think and talk about light is not greatly changed by the optical discovery, nor is there much reason why it should be. Novelists can continue to write as they would have done before: "As the first sunbeams lit the snow-capped peaks, and the peach-coloured glow spread down the mountain-side, chasing away the shadows and restoring their colour to the sleeping villages below, Charles awoke with a groan." Nor need our everyday instructions be rephrased: "Keep this bottle away from strong light" need not be replaced by "Do not allow light of high energy-density to be propagated on to this bottle."

Something, however, would be lost if we never did as the physicist recommends. There is a familiar sense in which we *must* accept the new picture of optical phenomena, for certain of the purposes of physics at any rate. And so far we have not seen what it is that obliges us to do so.

To see this, we must examine in more detail how the Principle of Rectilinear Propagation enters into a physicist's explanations: only a close examination will show us clearly where it comes in. For the physicist will say, fairly, that the reason why we *must* regard shadows in the way he recommends is that only in this way can their occurrence and movement be explained: it is in the service of his explanation that the principle, and with it the new way of thinking about shadow-casting and the like, are to be accepted.

Consider, then, a specific situation of the kind in which the physicist will be interested: notice how he sets about explaining an optical phenomenon, and in particular where the principle comes into his account. Suppose therefore that the sun, from an angle of elevation of 30°, is shining directly on to a six-foot-high wall, casting a shadow ten and a half feet deep on the level ground behind the wall. Why, we may ask, do we find that the shadow is just ten and a half feet deep: why not fifty feet, or two? How are we to explain this fact?

"Well, that's easy enough," the physicist will say. "Light travels in straight lines, so the depth of the shadow cast by a wall on which the sun is directly shining depends solely on the height of the wall and the angle of elevation of the sun. If the wall is six feet high and the angle of elevation of the sun is 30°, the shadow *must* be ten and a half feet deep. In the case described, it just follows from the Principle of the Rectilinear Propagation of Light that the depth of the shadow must be what it is."

Now we must not jump to conclusions about the logical form of this explanation. We must ask, first, how it can be said to follow from anything that the depth of a shadow must be just ten feet six and nothing else. What sort of inference, what sort of following is this? Not a bare inference from one straightforward matter-of-fact to a different one, for, as Hume rightly insisted, there can be no 'must' about any such inference— only a 'usually does'. Not a deduction from a generalization to an instance either for, considered as a generalization, the principle is just not true: in diffraction, refraction and scattering light ceases to travel in straight lines. Further, there is nothing in the principle about *all* shadows being ten feet six ins. deep, rather than fifty feet or two feet, so the only inference of a syllogistic kind one could look for would be "All light travels in straight lines; what we have here is light; so what we have here travels in a straight line", and this leaves the substantial step unaccounted for. In any case, if the inference were of a syllogistic kind, it would be open to the objection that logicians have always said it was, that of circularity— since one would be justified in saying only, "Light always has travelled in straight lines; what we have here is light; so what we have here will almost certainly travel in a straight line". Somehow none of the kinds of inference we are accustomed to from the logic-books seems to fit the case.

This should not surprise us. The fact of the matter is that we are faced here with *a novel method of drawing physical inferences*—one which the writers of books on logic have not recognized for what it is. The new way of regarding optical phenomena brings with it a fresh way of drawing inferences about optical phenomena.

This will become evident if we look and see what a physicist does when asked to set his explanation out in more detail, and make its form explicit. For the natural thing for him to do at this point will be to draw a diagram: in this diagram, the ground will be represented by a horizontal line, the wall by

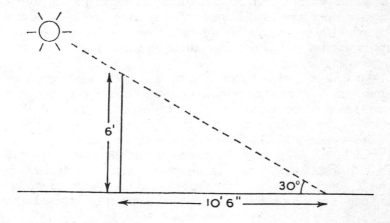

a vertical line, and a third line will be added at 30° to the horizontal, touching the top of the line representing the wall, and intersecting that representing the ground. This diagram plays a logically indispensable part in his explanation.

"Here", says our physicist, pointing to the third line, "we have the bottom ray of light which can get past the wall without being cut off. All the lower ones are intercepted, which explains why the ground behind the wall is in shadow. And if you measure the depth of the shadow on the diagram, you'll find that it is one and three-quarter times the height of the wall: that is to say, if the wall is six feet high, the shadow must be ten feet six deep."

Given the height of the wall and the sun, the physicist is in a position to discover by these means what depth the shadow of the wall will have; but he is able to do so only because he accepts the new account of optical phenomena and the inferring techniques that come with it. The view of optical phenomena as consequences of something travelling and the diagram-drawing techniques of geometrical optics are introduced hand-in-hand: to say that we *must* regard light as travelling is to say that only if we do so can we use these techniques to account for the phenomena being as they are. Neither the everyday nor the ancient way of talking and thinking about 'light' and 'sight' will make sense of the geometrical method of representing optical phenomena. And if the novel techniques of inference-drawing here used have not been recognized by logicians for what they are, that is probably because in geometrical optics one learns to draw inferences, not in verbal terms, but by drawing *lines*.

Of course, the fact that our physicist draws his diagram exactly as we have supposed, or draws any diagram rather than resorting to trigonometry, may not be important. But resort to some mathematical symbolism or other representational device is essential. As for the question, how the physicist's Principle of Rectilinear Propagation enables him to argue from the conditions of the phenomenon—the height of the wall and the angle of elevation of the sun—to his conclusion about the depth of the shadow: it does this, in practice, through the part it plays in the *representation* of the phenomenon concerned. In such a case as this, appeal to the principle means to the physicist something like the following—that the optical phenomena to be expected in this situation can be represented and so explained by drawing a straight line at the appropriate angle to the line representing the wall; that this line will mark the boundary between light and shade; and that one can read off such things as the depth of the shadow from the resulting diagram, confident that the result will be found to agree with observation within limits of accuracy greater than at present interest us.

The particular example here chosen may seem trivial, especially as we are limiting ourselves for the moment to circumstances in which there are no complicating phenomena such as refraction; but the steps we have gone through are of the very stuff of geometrical optics, and so in miniature of the exact sciences generally. Two things about it are worth noticing. First, the importance for physics of such a principle as that of the Rectilinear Propagation of Light comes from the fact that, over a wide range of circumstances, it has been found that one may confidently represent optical phenomena in this sort of way. The man who comes to understand such a principle is not just presented with the bare form of words, for these we have already seen to be on a naïve interpretation quite false: he learns rather what to do when appealing to the principle—in what circumstances and in what manner to draw diagrams or perform calculations which will account for optical phenomena, what kind of diagram to draw, or calculation to perform, in any particular case, and how to read off from it the information he requires.

Secondly, when a physicist has once drawn such a diagram of the 'optical state of affairs', he can use it not only to explain the original phenomenon, namely, the fact that the shadow was ten feet six ins. deep, but also to answer any number of other questions. It may, for instance, be asked what depth the shadow of our wall will have at a height of four feet from the ground. A horizontal line drawn two-thirds of the way up the line representing the wall intersects the line representing the 'light-ray' three and a half units out: answer, 3 feet 6 ins. Again, suppose that later in the year the sun shines directly on to the wall from an angle of 15°, instead of 30°. What will the depth of the shadow be then? A fresh line drawn at 15° to the

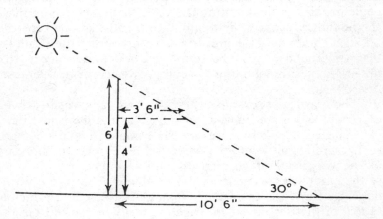

horizontal will cut the ground-line about thirty units from the wall-line: answer, about thirty feet. There is no limit to the number of such questions which a single ray-diagram can be used to answer.

2.3 *Inferring techniques and models are the core of discoveries*

At this point we can reconsider the question from which we began: the question what such a discovery as that light travels in straight lines amounts to. For we can see now that a vital part of the discovery is the very possibility of drawing 'pictures' of the optical state-of-affairs to be expected in given circumstances—or rather, the possibility of drawing them in a way that *fits the facts.*

Two things need saying to qualify this statement. To begin with, it is not necessary that the particular techniques we are here concerned with should be applicable in all circumstances. The way shadows fall and move, the patterns of light and shade cast by lamps, the places from which lights are visible or eclipsed

—it is enough that all these things can be accounted for, over a wide range of circumstances, in the way we have been studying. If under some circumstances refraction, diffraction and other such phenomena limit the use of these techniques, or require them to be supplemented, that does not destroy their value within the wide region to which they are applicable. Secondly, what is or is not to count as 'fitting the facts' has to be decided: there must be standards of accuracy. It can always be asked to what degree of accuracy a given method of representation can be used to account for a set of phenomena; and the best that we need demand of a theory is that it should fit the facts to as high a degree of accuracy as we yet have the means of measuring.

If these qualifications are borne in mind, we can answer our original question. The discovery that light travels in straight lines—the transition from the state-of-affairs in which this was not known to that in which it was known—was a double one: it comprised the development of a technique for representing optical phenomena which was found to fit a wide range of facts, and the adoption along with this technique of a new model, a new way of regarding these phenomena, and of understanding why they are as they are.

These are the core of the discovery. Compared with them, the particular words in which the discovery is expressed are a superficial matter: whether we speak of light as travelling or as being propagated is hardly important, for either is an equally good interpretation of the geometrical picture—at this stage, only so much of each notion matters as is common to them both. Further, the very notions in terms of which we state the discovery, and thereafter talk about the phenomena, draw their life largely from the techniques we employ. The notion of a light-ray, for instance, has its roots as deeply in the diagrams which we use to represent optical phenomena as in the phenomena themselves: one might describe it as our device for reading the straight lines of our optical diagrams into the phenomena. We do not *find* light atomized into individual rays: we *represent* it as consisting of such rays.

As for the Principle of the Rectilinear Propagation of

Light, the doctrine that light travels in straight lines, which figures in our sample explanation: we are now in a position to reconsider its status. We saw from the start that it could not be regarded as an empirical generalization of the kind logicians have so often discussed, since when so interpreted it is untrue. By itself, the principle tells us no additional facts over and above the phenomena it is introduced to explain, and if read as a factual generalization it would have to be qualified by some such clause as 'in general' or 'other things being equal' or 'except when it doesn't.' The point of the doctrine is in fact quite otherwise: its acceptance marks the introduction of the explanatory techniques which go to make up geometrical optics, namely, the model of light as something travelling from the source to the illuminated objects and the use of geometrical diagrams to infer what phenomena are to be expected in any given circumstances.

The doctrine is, so to speak, parasitic on these techniques: separated from them it tells us nothing, and will be either unintelligible or else misleading. For, as a discovery, it is opposed neither to the hypothesis that nothing is travelling, nor to the hypothesis that light is travelling differently—in both of which hypotheses the term 'travelling' must already have a sense. It is opposed rather to the use of a completely different model: to our thinking of optical phenomena for purposes of physics in wholly different terms—for instance, in terms of antennae from the eye seizing on the properties of the object—opposed, that is to say, to ways of thinking about light such that to talk of light travelling would not even be in place, ways which would lead us to formulate quite different questions and hypotheses about optical phenomena, in fact different *kinds* of question and hypothesis. As such, one might almost as well call the principle a 'law of our method of representation' as a 'law of nature': its role is to be the keystone of geometrical optics, holding together the phenomena which can be explained by that branch of science and the symbolism which, when interpreted in the way suggested by the model, is used by physicists to account for these phenomena.

2.4 *The place of mathematics and of models in physics*

How far are the things we have found in this particular example peculiar to it, and how far are they characteristic of discovery and explanation in the physical sciences generally?

In many respects the sample will be seen to be representative, once its extreme simplicity is allowed for. For in every branch of the physical sciences, the questions we have come to ask here can be asked again. Each branch is developed in order to account for a range of physical phenomena, and in each we can ask about the methods of representation and the models employed in doing so.

(i) Consider first the phenomena accounted for. In the case we have looked at, these will be such things as the changes in the distribution of light and shade as the sun moves across the sky, the times of eclipses and so on. But, as it stands, the range of the new principle is bounded. Any one branch of physics, and more particularly any one theory or law, has only a limited scope: that is to say, only a limited range of phenomena can be explained using that theory, and a great deal of what a physicist must learn in the course of his training is concerned with the scopes of different theories and laws. It always has to be remembered that the scope of a law or principle is not itself written into it, but is something which is learnt by scientists in coming to understand the theory in which it figures. Indeed, this scope is something which further research is always liable to, and continually does modify; and it is a measure of economy, apart from anything else, to state theories and laws in a manner which does not need to be changed whenever a fresh application of them is encountered.

(ii) Second, we have to consider the techniques of representation employed in the different branches of physics. In our sample, we are concerned solely with primitive mathematical techniques of a geometrical kind, including constructions with ruler and pencil and, at the most refined, the use of trigonometrical tables. It is from these techniques that this branch of optics gets its name of 'geometrical' optics. In it, we deal with optical phenomena by the use of geometrical pictures

—pictures in which straight lines represent the paths along which light is to be thought of as travelling—and work out rules for manipulating the straight lines of our figures so that they shall reflect as far as possible the observed behaviour of light, i.e. the optical phenomena concerned.

In some respects, our example is not characteristic, since the method by which the problems are handled is more than usually pictorial, giving the physicist what we have in fact called a 'picture' of the optical state-of-affairs. This vividness will make the example especially intelligible to the non-mathematician, but should not be allowed to mislead. For, though one can speak of this diagram as a picture, it is as well to remind oneself that such a picture would never appear in an art exhibition, however representational the tastes of the Hanging Committee—there being more than one kind of representation. The physicist's diagram is not valued for what the man-in-the-street would regard as a likeness, since the physicist's notion of light departs in important respects from the everyday one: still less is it valued on aesthetic grounds. Its point is a more prosaic one, that by the use of diagrams of this kind it has been found possible to show, and so explain, over a wide range of circumstances and to a high degree of accuracy, what optical phenomena are to be expected.

Wherever possible, physicists would like to be able to represent the phenomena they are studying pictorially: where this is possible, one can 'see' the force of their explanations in a specially convincing way. For the same reason, geometry seemed to mathematicians in the seventeenth century to be superior to algebra: they felt that algebra provided only a short-cut to truths which geometry displayed. But this can rarely be done to anything like the degree to which it can in geometrical optics. Only in a very few branches of physics does the drawing of diagrams play a logically central part. In most branches the logical role played in geometrical optics by diagrammatic techniques is taken over by other less primitive kinds of mathematics; and these are often of a complexity and sophistication far greater than could ever be handled diagrammatically. Yet however sophisticated and complex these may become, they play a part comparable to that of picture-drawing in geometrical optics: they serve, that is to say, as techniques of inference-drawing. In dynamics, for instance, the counterparts of our geometrical diagram are the equations of motion of the system of bodies under investigation. Given a suitable description of a system, a physicist who has learnt Newtonian dynamics will be in a position to write down its equations of motion: these equations can then be thought of as providing, in a mathematical form, a 'picture' of the motions of the system, logically parallel to that which our diagram gives of optical phenomena. Using the equations, he will be able to compute such things as the velocity a particular body will have when it has risen to such-and-such a height from the ground, and the height at which it will cease to rise; just as, from our diagram, we can discover the depth of the wall's shadow at different heights from the ground.

This is a point worth emphasizing, for the place of mathematics in the physical sciences is something people tend to find mystifying. One is even told at times that physicists work in two worlds, the 'world of facts' and the 'world of mathematics', which makes one wonder how it can be that the world around us is, as they imply, interpenetrated by this other, unseen 'mathematical world'. But there is no point in talking about a separate 'world of mathematics', unless to remind ourselves not to look for all the features of, e.g., light-rays in sunbeams and shadows alone; the world in which our theoretical concepts belong being as much the paper on which our computations are performed as the laboratory in which our experiments are conducted. If mathematics has so large a place in the physical sciences nowadays, the reason is simple: it is that all such complex sets of exact inferring-techniques as we have need of in physics can be, and tend to be, cast in a mathematical form.

Certainly none of the substantial inferences that one comes across in the physical sciences is of a syllogistic type. This is because, in the physical sciences, we are not seriously interested in enumerating the common properties of sets of objects, but are concerned with relations of other kinds. This point will be taken up again later, when we consider the differences between the physical sciences and natural history. The operations we perform and the observations we make in physics are not just head-counting; the logical form of the conclusions we reach is not that of a simple generalization; and the kinds of inference we can draw as a result are not syllogistic inferences. Indeed, the inferences of physics are substantial just because they are so much more than transformations of our observation-reports. If one has counted over all As and checked that they are all Bs, one has thereby checked that any particular A one selects will be a B: subsequent inferences from "All As are Bs" to "This A is a B" are automatic. On the other hand, if one has measured the height of a wall and the angle of elevation of the sun, one has not thereby measured the depth of the shadow cast by the wall: yet this is something which the techniques of geometrical optics enable one to infer, providing the circumstances are of a kind in which physicists have found the techniques reliable.

The same is true more generally. The heart of all major discoveries in the physical sciences is the discovery of novel methods of representation, and so of fresh techniques by which inferences can be drawn—and drawn in ways which fit the phenomena under investigation. The models we use in physical theories, which tend to be featured in popular accounts as though they were the whole of the theories, are of value to physicists primarily as ways of interpreting these inferring techniques, and so of putting flesh on the mathematical skeleton. The geometrical diagram used in our optical example is lifeless unless we think of light as something travelling 'down the dotted line': only so shall we be able to see how it is that the diagram explains the phenomena it does. But equally the model of light travelling, remote as it is from our non-scientific way of thinking about light and shade, is pointless without the diagram. To present a theory simply in terms of the models employed is to forget the thing that matters above all, and to leave the *use* of the model completely unexplained.

In practice, then, a theory is felt to be entirely satisfactory only if the mathematical calculus is supplemented by an intelligible model. It is not enough that one should have ways of arguing from the circumstances of any phenomenon to its characteristics, or *vice versa*: the mathematical theory may be an excellent way of expressing the relations we study, but to understand them—to 'see the connection' between sun-height and shadow-depth, say—one must have also some clearly intelligible way of conceiving the physical systems we study. This is the primary task of models: for know-how and understanding both mathematics and models are wanted. The impossibility of providing a single model by which to interpret the mathematical theories of quantum mechanics has accordingly been felt by many to be a drawback—and even spoken of, frivolously or confusedly, as showing that 'God must be a mathematician'. Previously, it had always been possible to match one inferring technique over its whole range of application with a single model: it is this which, for demonstrable reasons, cannot be done in the case of quantum mechanics, so that while a wave-model may be of use in some applications of the theory a particle-model is more suitable in others.

(iii) Let us next look at the notion of a model a little more closely. Consider once again our example: in that sample explanation, the diagram provides, as we have seen, something in the nature of a picture of the optical state-of-affairs; a picture with the help of which we can infer things about the shadows and other optical phenomena to be observed under the circumstances specified. But to understand how the explanation works, it is not enough to point to the phenomena on the one hand and the physicist's diagram on the other. For the physicist uses other terms, having at first sight nothing to do either with shadows or with diagrams, which nevertheless constitute in some ways the heart of the explanation. He talks, for instance, of light 'travelling', of rays of light 'getting past the wall' or 'being intercepted by it', and declares that this interception of light by the wall is what—fundamentally—explains the existence of the shadow.

A point which we made earlier is worth repeating here. In developing geometrical optics, we have passed from regarding the phenomena of light and shade as primitive phenomena, which have just to be accepted and left unexplained, to seeing them as the common effects of something, for which 'light' is again the word we use, travelling from the sun to the objects lit by it. This step means coming to speak and think about the phenomena in a new way, asking questions which before would have been unintelligible, and using all the words in our ex-planations—'light', 'travel', 'propagated', 'intercept' and the rest—in quite novel and extended senses. Later on, of course, we come to feel that these are the most natural extensions in the world; so much so, in fact, that we forget that they ever had to be made.

Since these uses of the words are extended ones, only some of the questions which ordinarily make sense of things we can describe as travelling are applicable to the novel traveller, the physicist's new entity, 'light'. Some of the questions which we do not ask in the new application are ones which anyone would feel to be obviously irrelevant, some of them are ones which in the everyday application are central. Thus we find it natural enough not to ask of 'light' whether it travels by road, rail or air, or whether it has a single or return ticket—though remember that the discredited 'ether' was meant in part as an answer to the question "By what means does light travel?" But it is stranger to find that nothing in geometrical optics gives us any occasion to discuss the question *what* it is that 'travels'. So far as geometrical optics is concerned, it is enough that we have as the gram-matical subject of our sentences the bare substantive 'light', and it does not matter whether or no we can say any more about it.

This point is worth following up. No doubt it is an import-ant feature of the new way of thinking about optics that we *are* prompted to ask such questions as "What travels?" There are indeed many phenomena in accounting for which we come to think of the grammatical subject as having a physical counterpart: these are the phenomena with which we are concerned in physical optics. Nevertheless, the questions with which physical and geometrical optics are concerned are logically independent. We know that light starts off from lamps, stars and other shining bodies, and ends up on illuminated surfaces: all we need ask, therefore, in geometrical optics are the questions, "Where from? Where to? And by what path?" The whole of geometrical optics could have been, and much in fact was developed, without there being real backing for any particular answer to the question "What is it that travels?" Even the question "How fast?" was answered by Römer in 1676 from observations on the eclipses of the satellites of Jupiter, before any substance had been given to the bare grammatical substantive 'light'.

This is something which one quite often finds in the physical sciences. At the stage at which a new model is introduced, the data that we have to go on, the phenomena which it is used to explain, do not justify us in prejudging, either way, which of the questions that must normally make sense when asked of things which, say, travel will eventually be given a meaning in the new theory also. The acceptance of the model is justified in the first place by the way in which it helps us to explain, represent and predict the phenomena under investigation. Which of the questions that it suggests will be fertile and what hypotheses will prove acceptable are things which can be found out only in the course of later research, in a manner which we shall have to examine later.

One might speak of models in physics as more or less 'deployed'. So long as we restrict ourselves to geometrical optics, the model of light as a substance travelling is deployed only to a small extent; but as we move into physical optics, exploring first the connexions between optical and electromagnetic phenomena, and later those between radiation and atomic structure, the model is continually further deployed.

The process by which, as we go along, fresh aspects of the model are exploited and fresh questions given a meaning is a complicated one, and one which needs to be studied in detail for each fresh branch of physical theory if the logic of that theory is to be clearly understood. At the moment, all we need to note is this: although only some of the questions which ordinarily apply to things which, e.g., travel do so in the extended use, one cannot say beforehand which questions will and which will not apply, and it has to be discovered as time goes on how far the old questions can be given a meaning in the new type of context. Some of the most important steps in physics have in fact consisted in giving to more of these questions interpretations they did not have before (e.g. the development of physical optics, and the introduction of the notion of sub-atomic structure); others in doing something which was in many ways more difficult to do, namely, giving up hope of answering questions which up to that time had seemed perfectly natural and legitimate (e.g. Leibniz on the mechanism of gravity, and the nineteenth-century disputes about the luminiferous ether).

The unlimited deployability of physical models seems to be one of the things Planck and Einstein have in mind when they insist that electrons and gravitational fields are as real as tables and chairs and omnibuses.[1] For there is no denying the differences, in logical status as well as in physical properties, between such theoretical entities and notions as 'electrons', 'genes', 'potential gradients' and 'fields', and everyday objects like buses and tables. But what physicists are entitled to insist is, that their models need not necessarily be spoken of, deprecatingly or otherwise, as theoretical fictions; for to regard them all equally as fictions would imply that there is no hope of deploying any of them very far, and would suggest that it was risky following up for any distance the questions which they prompt us to ask.

This would be a great mistake. It is in fact a great virtue of a good model that it does suggest further questions, taking us beyond the phenomena from which we began, and tempts us to formulate hypotheses which turn out to be experimentally fertile. Thus the model of light as a substance in motion is a good model, not only because it provides us with an easily intelligible interpretation of the diagrams of geometrical optics —though this is a *sine qua non*—but also because it carries us beyond the bare picture of something unspecified travelling, no matter what, and leads us to speculate about light-particles or light-waves as the things which travel, or are propagated:

[1] This topic will be taken up in more detail in Sec. 4.7 below.

these speculations have borne fruit. Correspondingly, the models of thermal and gravitational phenomena as the effects of caloric and gravitational fluids were bad models, since the questions they prompt one to ask turned out in fact to be as unprofitable as those which the Greek antennae-model led one to ask in optics.

Certainly it is this suggestiveness, and systematic deployability, that make a good model something more than a simple metaphor. When, for instance, we say that someone's eyes swept the horizon, the ancient model of vision as the action of antennae from the eye is preserved in our speech as a metaphor; but when we talk of light travelling our figure of speech is more than a metaphor. Consequently, when people say that to talk of light travelling in some sense reflects the nature of the world in a way in which to talk of eyes as sweeping the horizon does not, they have some justification. For to say that "Light travels" reflects the nature of reality, in a way in which "His eyes swept the horizon" does not, is to point to the fact that the latter remains *at best* a metaphor. The optical theory from which it came is dead. Questions like "What sort of broom do eyes sweep with?" and "What are the antennae made of?" can be asked only frivolously. The former does more: it can both take its place at the heart of a fruitful theory and suggest to us further questions, many of which can be given a sense in a way in which the questions suggested by "His eyes swept the horizon" never could.

2.5 *Theories and observations are not deductively connected*

One can, therefore, afford to speak of physical theories as drawing their life from the phenomena they are used to explain. If the layman is told only that matter consists of discrete particles, or that heat is a form of motion, or that the Universe is expanding, he is told nothing—or rather, less than nothing. If he were given a clear idea of the sorts of inferring techniques the atomic model of matter, or the kinetic model for thermal phenomena, or the spherical model of the Universe is used to interpret, he might be on the road to understanding; but without this he is inevitably led into a *cul-de-sac*.

It is as though we were to show a brand-new gas geyser, still lying in its box, to a man who was quite unfamiliar with the mechanical devices of Western life, and were to say to him, "That heats water". We should have no right to be surprised if he thought that we were showing him a robot cook—this is the counterpart of the Man Friday fallacy. The least we can do for him is to say, more lucidly, "This is something which can be used for the heating of water", and indicate roughly the way in which it would have to be assembled in order to do what it was designed to do. The sentence "This heats water", uttered in such a context, is a condensed form of words intelligible only to those familiar with the kind of device in question. No geyser heats water or anything else so long as it is left lying in its box surrounded by shavings: it must be joined up to the mains in the way the makers specify before there is any hope of it doing its job. The same holds of sentences like "The atomic model explains all known chemical phenomena". Once again, the atomic model by itself can do nothing at all; but it can be used, in the way in which it was designed to be used, in explaining the changes and processes that chemists study. As for "Heat is a form of motion", this leaves almost everything unsaid. Light, as we ordinarily understand the word, is not something which can be spoken of as travelling: so too, heat is no more a form of motion than dampness is a form of departure.

One philosopher of science who saw the importance of this point was Ernst Mach. He, too, used to insist that the adoption of new theories and models was justified only by the observational and experimental results which led up to them; but he overstated his case in an interesting way. For the conclusion he came to was that the statements of theoretical physics were abridged descriptions of the experimental results, comprehensive and condensed reports on our observations, and nothing more. He thought that we should be justified in accepting our theoretical conclusions only if these were logically constructed out of the reports of our experiments; that is, related to them in a deductive way, as strictly as statements about 'the average Englishman' and data about individual

Englishmen. Only in this way, he concluded, could one avoid either anthropomorphism or what we have called 'the Man Friday fallacy'. All talk about explanation, especially in terms of 'insight into causal connexions', seemed to him to run into these difficulties: causal connexions were in his view as mythical as the personage, Light, whom a complete novice might suppose us to regard as 'responsible for making shadows'.

The confusion of thought which led Mach and the Phenomenalist School to this conclusion is not entirely easy to sort out, and we shall have to return to the matter in later chapters. But it is essential to see at the outset that there can be no question of observation-reports and theoretical doctrines being connected in the way Mach thought: the logical relation between them cannot be a deductive one. This comes out clearly from our example: however many statements you collect of the form, "When the sun was at 30° and the wall six feet high, the shadow was ten feet six ins. deep", you will not be able to demonstrate from them in a deductive manner the necessity of the conclusion, "*Ergo*, light travels in straight lines". Not that this is anything to worry about; for, given on the one hand statements about everyday things, like lamps, the sun, shadows and walls, and, on the other hand, theoretical statements in terms of the physicist's concept, light, how can we even imagine finding deductive connexions between them? The types of sentence which are deductively related are always taken out of roughly the same drawer and stated in similar terms— for instance, "Fish are vertebrates", "Mullet are fish" and "Mullet are vertebrates". But the two classes of sentence now under consideration are stated in quite dissimilar terms, and in them language is being used in radically different ways.

To say "Light travels in straight lines" is, therefore, not just to sum up compactly the observed facts about shadows and lamps: it is to put forward a new way of looking at the phenomena, with the help of which we can make sense of the observed facts about lamps and shadows. But this is not the same as to say, "One can represent the phenomena thus: . . .", or "Physicists now regard light and shade thus: . . .". Rather it is to *play the physicist*, to speak the words of one who regards them in the new way. In view of this, we can see how misleading it might be to say, without qualification, that "Light travels in straight lines" is a law as much of our method of representation as of nature. For the discovery that light travels in straight lines was certainly not a discovery about *physicists*—i.e. that they can choose, or do choose, to represent optical phenomena in a geometrical manner. Not at all: if they did not mind about the consequences, they could choose to represent them anyhow they pleased. There is an additional discovery, beyond the fact that they do so choose, which alone shows the importance of the principle for physics: namely, that if one does so represent them, it is possible to explain optical phenomena of a wide range of types—light and shade, eclipses and so on—with certain restrictions (no refraction etc.) but to a high degree of accuracy; and further, as we shall see, that these techniques can, with the aid of simple rules, be extended to situations involving refraction and reflection and other phenomena so far ruled out.

Still, the difficulty Mach felt is one that we are all liable to feel when we first notice the logical differences between theoretical statements, like "Light travels in straight lines", and observation-reports like "The shadow was ten feet six ins. deep". It is natural for a logician to suppose that, in order to justify a theoretical conclusion, one must collect sufficient experimental material to entail it; and that, if one does anything less, the theoretical conclusion will assert something more than the experimental data warrant. Mach, at any rate, was very keen to show that laws of nature 'contain nothing more than' the facts of observation for which they account. But this is a mistake. For it is not that our theoretical statements ought to be entailed by the data, but fail to be, and so assert things the data do not warrant: they neither could be nor need to be entailed by them, being neither generalizations from them nor other logical constructs out of them, but rather principles in accordance with which we can make inferences about phenomena. This point will be made clearer in the next chapter. To justify the conclusion that light travels in straight lines, we do not have to make observations which *entail* this conclusion: what we have to do is to show how the data we have can be accounted for *in terms of* this principle. The absence in this case of a deductive connexion is not to be thought of as a *lack* of connexion, any more than a hammer need be thought of as lacking a screw-thread: justification here calls for something other than a demonstrative proof.

The real difficulty is to avoid stating the obvious in a misleading way. Einstein, for instance, objects to Mach's doctrine but almost tips over backwards in his effort to rebut it: he speaks of physical theories as 'free products' of the human imagination. Granted that discoveries in theoretical physics are not such things as could be established either by deductive argument from the experimental data alone, or by the type of logic-book 'induction' on which philosophers have so often concentrated, or indeed by any method for which formal rules could be given.[1] Granted that discoveries in the physical sciences consist in the introduction of fresh ways of looking at phenomena and in the application of new modes of representation, rather than in the discovery of new generalizations. Perhaps, too, the recognition of fresh and profitable ways of regarding phenomena is, in part at least, a task for the imagination, so that Einstein can say of them, as he says of the axiomatic basis of theoretical physics, that they "cannot be abstracted from experience but must be freely invented. . . . Experience may suggest the appropriate [models and] mathematical concepts, but they most certainly cannot be deduced from it." But we must not be tempted to go too far. This is not work for the untutored imagination. It may be an art, but it is one whose exercise requires a stiff training. Though there is nothing to tell just what new types of model and mode of representation scientists may not in time find it profitable to adopt, nor any formal rules which can be demanded for discovering profitable new theories, theoretical physicists have to be taught their trade and cannot afford to proceed by genius alone.

[1]This is why it is so unfortunate that logicians have come to speak of scientific discovery as 'inductive inference': where no rule of inference could ever be given, the very notion of inference loses its point. Discovery is, rather, a prerequisite of inference, since it includes the introduction of novel techniques of inference-drawing.

One cannot teach a man to be imaginative; but there are certain kinds of imagination which only a man with a particular training can exercise.

The situation is rather like that in which, as we are sometimes told, unbreakable glass or saccharin or radio-activity or blotting-paper was discovered 'by accident'. Again this is a misleading way to talk: such discoveries are not made by accident, even though they may be made *as a result of* an accident. Most people, if they knocked a glass jar on to a stone floor and it did not break, would pick it up, thank their lucky stars and leave things at that: only a scientist with the right initial training would know just how odd a happening this was, and would be equipped to find out what had happened to the jar beforehand that prevented it from shattering. It might be a piece of luck that one scientist rather than another first noticed the phenomenon; but it would not be luck which guided the rest of his investigation. It may, likewise, be a fertile imagination which first leads one physicist rather than another to explore the possibilities of some novel theory; but again, it is trained skill quite as much as imagination which guides him in the exploration once it is begun.

2.6 *Physics is not the natural history of the inert*

There is one final point about the sorts of things which count as discoveries in the physical sciences which must be emphasized at the outset: this will help us to understand the differences between explanatory sciences, such as physics, and descriptive sciences, such as natural history. The point can be put concisely by saying: physicists do not hunt out regularities in phenomena, but investigate the form of regularities whose existence is already recognized. As it stands, this may seem rather a dark saying; so let us take another look at some examples.

It must have been recognized that there was some regularity in the way in which shadows were cast long before this fact was scientifically explained: the development of geometrical optics made clear and explicit the nature of a regularity which had previously been appreciated only roughly. Again, it was known that the planets moved in a regular way, and these regularities had been studied, for many centuries before there was any dynamical theory to make sense of them: the development of dynamics once again made intelligible regularities whose existence was previously known, but whose exact nature and limits had not been understood.

The consequences can be seen if one looks at the starting-point of the physical sciences, and at the scientist's opening moves. For the regularities of everyday experience, with which we are all familiar, provide him with a natural point of attack: and the questions he will begin by asking are not *"Are there laws of motion, optics or chemical combination?"* but "What are the *forms* of these laws?" With such a starting-point, one question does not need to be asked: namely, *whether* there is any connexion between, say, the slope of a hill and the way a stone moves when placed on it, or between the position of the sun in the sky and the length of shadows. Like the rest of us, the scientist knows very well that these things are, in some way to be discovered, interdependent: the form of his first question will therefore be not *"Are* these things interdependent?" but *"How* do they depend on one another?"

Philosophers have sometimes talked as though science could be divorced from common experience, and as though the scientist had a completely free choice of starting-point. Now it is true that, once his subject is established, a scientist will choose what experiments to perform and how to perform them on the basis of scientific considerations alone—we shall see later how closely the conditions of an experiment are determined by the nature of the theoretical problem on which the experiment is designed to throw light. But it does not follow that, at the very beginning of a science, the investigator can start just anywhere. Though we can hardly speak of the ordinary man having theories about natural phenomena, it is nevertheless such everyday regularities as we have been concerned with in our optical example, and the departures from them, that pose to the scientist his first theoretical problems.

To point to the very beginning of a science is, in fact, to make an artificial division. Current theoretical problems in, say, the dynamics of fast-moving particles arise out of the limitations of the Newtonian theory; the Newtonian theory of motion was the solution of problems posed by the limitations of the Aristotelian theory, since it was the failure of Aristotle's dynamics to deal with acceleration that focused attention on that phenomenon during the sixteenth and seventeenth centuries; Aristotle's dynamics in its turn was an attempt to systematize and extend our ordinary ideas about motion; and where exactly in this sequence are we going to draw the line? At each stage, the centre of interest depends on the current background of ideas about motion. These provide the standard of what is normal, of what is to be expected, and it is primarily departures from this standard which are spoken of as 'phenomena', that is, as happenings requiring explanation. When we go back to the stage in any science at which the first systematic attempts were made to theorize, to connect up the phenomena in that field, it is the notions of contemporary common sense which provide the background of ideas by reference to which phenomena are chosen for investigation. And, since common sense in this context means 'recognizing the regularities with which we are familiar from everyday experience,' it is natural that these should play a prominent part in the early stages of most of the sciences—so that it was, for instance, from a study of breathing and burning ('respiration' and 'combustion') that the *savants* of the late eighteenth century first began to understand the nature of chemical reactions, and gave Dalton his chance to make of chemistry something more than a collection of isolated industrial techniques and conjuring tricks.

From this we can see the source of one of the differences between the physical sciences and natural history. In physics we cannot afford to begin where we like. Rather, as Newton puts it, we must trace out the laws from the phenomena in a few simple cases; and apply what we discover in these cases, as principles, when we turn to more involved cases. "It would be endless and impossible to bring every particular to direct and immediate observation"; so the physicist only has time to investigate in detail the behaviour of the simpler systems.

If you bring a physicist or chemist a box containing an unidentified assemblage of things, he may be perfectly entitled to brush aside your request to be told how it works and what will happen if you do different things to it: the contents of your box will probably not be a suitable object of study. He may possibly, given time, discover what it is that you have brought him, and so be able to answer your questions—at any rate, in certain respects, and to a limited degree of accuracy. But unless the assemblage is a particularly simple one, the task of identification will be lengthy, and the scientist will be within his rights if he regards you as having interrupted, not contributed to, the progress of his work.

In natural history, things are quite otherwise. Whatever kind of living creature we come across, it will be equally fair to ask the naturalists what it is, and what its habits are. Any kind of animal is a 'suitable object of study' for the natural historian; and if at a particular stage in history one species has been studied more than others that will not be for theoretical reasons, but for practical ones—for instance, because it is easy to feed and is not afraid of humans, so that it can be watched without the need for elaborate hides. All living creatures equally may be subjects for the natural historian, but, for theoretical as well as practical reasons, observation and experiment in the physical sciences have to be highly selective.

This, however, is a comparatively minor difference between the descriptive and explanatory sciences. The larger differences have a more subtle origin, and we must try to get clear about it. Notice for a start, then, that the kinds of regularity we encounter in everyday life, which form the starting-points of the physical sciences, are hardly ever *invariable*; and correspondingly, the degree of system in everyday language is limited. Only rarely can one infer from an everyday description of the circumstances of a phenomenon just what form it will take.

Some small amount of system there is, reflecting the familiar regularities that every child soon discovers. This is most clearly to be seen in the use we make of law-like statements: "Don't hit the window: glass is brittle (i.e. breaks if hit)". But this system is not particularly reliable. All such inferences in ordinary language are open to qualification: "This is made of wood, so it *must* float—unless it's *lignum vitae* or is water-logged", "You can see the road's straight, so that *must* be the shortest way—unless we're up against some optical illusion". These inferences depend on physical or natural-historical regularities of whose scope we have only a vague idea, and they are therefore liable to exceptions. We should not be very much surprised, e.g., to find another kind of wood besides *lignum vitae* which refused to float.

Many of the delights of childhood, indeed, consist in defeating these regularities. It may be fun to roll a stone down-hill; but it is much more fun to fill a balloon with gas, and watch it float up to the ceiling. We only expect these regularities to hold on the whole, and we are not particularly disconcerted when we encounter the exceptional case.

Nor need these limitations matter for most practical purposes. A carpenter need be no physicist to know that, in the main, the way two planks look is a good guide to the way they will fit, and that if the foot of a plank is in water the look of it will no longer be such a good guide. The ability to explain *why* a plank looks bent in water would not simplify his tasks as a carpenter: his professional attitude to this phenomenon will accordingly be one of indifference. So long as he is able to tell in practice when look will and will not be a good indication of fit, he need not be particularly interested in the optical theories required to explain these facts.

It is the mark of the physical scientist, on the other hand, to be interested in such regularities and their limitations for their own sakes. It is a matter of professional concern to him to find out what exactly they amount to, why they hold and fail to hold when they do, under what conditions departures are and are not to be expected—and, if possible, to develop a theory which will explain all these things. The questions which are of importance to him are, accordingly, these: "What form does the regularity take, in the cases in which it occurs?" and "Under what circumstances are we to expect it to occur?" To put the point briefly, the physicist seeks the *form* and the *scope* of regularities which are found to happen, not universally, but at most on the whole.

This point has been consistently misunderstood in text-book discussions of scientific method. Starting with a study of the syllogism, the probability calculus and the calculus of classes, and then coming to the physical sciences, logicians have been misled by their earlier preoccupations and interests, vested as they are in formal systems of considerable refinement and elaboration, into looking for the wrong things. One form of statement alone has commonly been examined, the universal empirical generalization; and only the more detailed treatments of the subject have even succeeded in passing on from "All As are Bs" to "The probability of an A being a B is 3/5" and "Conditions C_1, C_2 and C_3 being fulfilled, all As are Bs." The consequences have been unfortunate. Laws of nature have been confounded with generalizations, such sentences as "All swans are white" and "All ravens are black", being gravely discussed under this heading. Hypotheses have been treated as though they were simply laws of which we are not yet confident, since they have not been checked in a sufficient number of instances. As for experiments, these have been presented as first cousins of the Gallup poll—concerned only with *how often* different pairs of properties are found to go together.

But to accept such an account is to treat physics as though it were a kind of natural history, and so to waste one's labour. Natural historians may be interested enough in discussing whether or no all ravens are black, and whether all mice like cheese. But so long as one remains within natural history there is little scope for *explaining* anything: "Chi-chi is black, because Chi-chi is a raven and all ravens are black" is hardly the kind of thing a scientist calls an explanation. Indeed, among scientists, to say that a newly fledged subject is still in 'the natural-history stage' is a way of depreciating it: natural history and the like are felt to lack many of the essential features of a full-grown science, and to be entitled to the name of sciences only conditionally and out of courtesy.

This practice is not entirely fair to natural history, since as soon as an observer suggests, e.g. how the colouring of some sub-species of rat may be explained in terms of its environment, he is promoted from 'natural historian' to the more respectable rank of 'zoologist'. But the feeling has some justification. For, if explanatory power is regarded as the stamp of a science, then the shallow explanations which are all that we can demand of natural history take us little beyond the point which, in dynamics, every child has reached: "This rolls downhill, because this is a stone, and stones generally do roll down hills." How different are the sorts of conclusion aimed at in the physical sciences: "Light travels in straight lines", "The hydrogen atom consists of one proton and one electron"— the very point of such statements lies in their explanatory fertility; and in hardly a single respect are they comparable with the generalizations about habits or plumage which are all that natural historians can announce.

2.7 The crucial differences between physics and natural history

The reason for the differences between generalizations about habits, plumage, etc. ('habit-statements') and what, by contrast, may be called 'nature-statements', will become evident as we go along. But there is one point of general importance that requires to be touched on here. This has to do with the question, what sorts of subject-matter the two types of statement can have—i.e. what sorts of grammatical subject they can contain. Here at last we shall begin to see how the logical differences between the two classes of statement spring from differences between the two kinds of scientific activity.

The subject-matter of the natural historian's habit-statements is the same as that of everyday speech and affairs: at most, the natural historian will sub-divide the everyday classification in ways we would not normally bother to do, distinguishing, for example, between the Spotted Woodpecker, *Dryobates major anglicus*, and the Northern Spotted Woodpecker, *Dryobates major major*. The task of identifying to what class a given subject belongs will not in general be a highly technical one; although there may be difficult or borderline cases which have to be left to the expert, in the main, as

Wittgenstein has remarked, 'what is or is not a cow is for the public to decide'.

Being tied, in its essentials, to the everyday classification, the natural historian has left to him to discover such things as what breeding-habits are common to all gulls, and what proportion of the North Sea herring shoals passes through the Straits of Dover in the average summer. In consequence, his conclusions are, from a logician's point of view, both quite straightforwardly factual and open to logical analysis in the traditional way: they will fit without appreciable distortion into the familiar patterns, "All As are Bs" "All As which are also Cs are Bs", "The proportion of As which are Bs is 3/5", and so on.

Furthermore, since the classification of his subject-matter is made along everyday lines, it is not open to the natural historian to modify its principles in the light of his discoveries. Were he to find that half the house-mice in England were herbivorous and half carnivorous, and that these two sets of mice did not interbreed, he could and would distinguish between the two classes and, if the circumstances made this appropriate, might come to speak of them as two different species of mouse; but he would not be at liberty to say either "One half lives on lettuces, so they can't be mice after all", or "Only the ones that live on lettuces are to be regarded as mice." Or rather, if he did insist on doing so, the agreement of the public would be a sign, not of his expert knowledge, but of his prestige—like the agreement never to call whales 'fish'.

When one turns from natural history, with its habit-statements, to the nature-statements of the physical sciences, one finds that the situation is markedly different. In talking about the phenomena they study, physicists need no more confine themselves to the everyday classification of the things they encounter than they do to the more elementary logical forms. Reclassification of subject-matter in the light of discovery is the rule in the physical sciences:[1] the decision, what is or is

[1] It is in this way, for instance, that the classification of kinds of matter by origin and the like, i.e. as 'wood', 'water', 'stone', etc., comes to be supplemented by the classification into kinds of chemical substance, as 'hydrogen', 'carbon dioxide', etc.

not to be spoken of as 'a purely gravitational phenomenon'—as opposed to 'a cow'—becomes therefore a highly technical matter, and the grounds on which it is made change as the theories of science develop.

This fact has important consequences for the logic of the things the physicist says. In natural history, one can distinguish sharply between two stages in any piece of research: the initial step of identifying an animal—unnecessary, of course, if it was bred in the laboratory—and the subsequent process of studying its habits. In the physical sciences, there is no such sharp division: the things that come to light as one goes along will frequently lead one to relabel the system being studied. The statement, "This can't be classified as a mouse, for it eats lettuce", may be inadmissible, but its physical counterpart is quite conceivable: "This can't be classified as a purely gravitational phenomenon, for the orbit is nutating as well as precessing." Now we can account for something we noticed earlier, namely, the impossibility of treating the statements of theoretical physics as universal empirical generalizations. The reason why the form "All As are Bs" does not fit the statements of physics is this: only where one can ask *separately*, first, "What are these?" (Answer: As), and then, "What common properties have they?" (Answer: being Bs), is "All As are Bs" the natural form in which to couch one's conclusions. One can make this separation in natural history; but in the physical sciences the two questions are interdependent, and in consequence the simple generalization is out of place.

What is the point of the physicist's reclassification? To see this, recall that it is his aim to find ways of inferring the characteristics of phenomena from a knowledge of their circumstances. This aim is one which ordinary language, being largely devoid of system, does not serve very well. To speak of something as a 'blackboard', for example, implies hardly anything about how it will behave. No doubt, if it explodes, or crumbles into dust, or vanishes without warning, we shall be very much surprised, and try to find an explanation; but it cannot be said to be *implied* by one's description of it as a blackboard that these things either will or will not happen to it, however unexpected and inconvenient they may be. If the manufacturers of blackboards found that their products could not be guaranteed against disintegration—in fact, that they were all liable to crumble away at an unpredictable time after manufacture, like radioactive nuclei—that would not stop us talking of them as blackboards, any more than the finite life of the filament stops us calling electric lamp-bulbs 'lamp-bulbs'. No reclassification or other linguistic steps would be called for. We should simply have to lay in a stock of replacements; and, if things became too bad, school-teachers would take care to say, "I'll leave this graph on the board, and we'll talk about it next time, *with luck*."

Once again, how different is the situation in the physical sciences. There the specification of a system carries rigorous implications about its behaviour. The chemist analysing a specimen, for example, will not be satisfied until he can account for its observed chemical properties by reference to its constitution as strictly as we accounted for the depth of the wall's shadow; and if two specimens, both from the same source, have quite different properties, he will not be satisfied to regard them as being of the same substance or as having the same structure. His classification must take account of the differences between their properties: if it does not allow for these differences, so much the worse for his classification. Indeed, the classification-system scientists employ changes as time goes on, and the way in which it does so shows what their ideal is: that, from a complete specification of the nature of any system they have under investigation, it should be possible to infer how it will behave, in as many respects and to as high a degree of accuracy as possible.

2.8 *Description and explanation in science*

Natural historians, then, look for regularities of given forms; but physicists seek the form of given regularities. In natural history, accordingly, the sheer accumulation of observations can have a value which in physics it could never have. This is one of the things which the sophisticated scientist holds against natural history: it is 'mere bug-hunting'—a matter of collection, rather than insight.

Now there is something important in this way of putting the difference, which is reflected in the sorts of thing that could be accepted as observations in physics and natural history respectively. As one cannot start doing physics just anywhere, so also there are very definite limits to what will count in physics as an observation. Gilbert White was able to make valuable contributions to natural history by keeping a diary of the things he noticed as he went around the Hampshire countryside, for in natural history all facts about fauna are logically on a par. But, as Popper has pointed out, one could not hope to contribute to physics in this way. However full a note-book one kept of the phenomena one came across in the ordinary course of one's life, it would in all probability be of no value to physicists at all. In physics, it is no use even beginning to look at things until you know exactly what you are looking for: observation has to be strictly controlled by reference to some particular theoretical problem. Just how close the connexion has to be, we shall see in the next chapter.

On this point, Mach and his followers again tend to be misleading. One finds them, for instance, identifying 'observations' on the one hand, and 'sense-data' on the other, which suggests that we are for ever making observations. This is a confusing practice, for it entangles the logical problems of physics with the philosophical problems which have to do with perception and material objects. Furthermore, it does this needlessly, since it is not difficult to keep the terms 'sensation' and 'observation' sorted out: as though one only had to open one's eyes to 'make observations'. This tendency is probably connected with Mach's desire to show that all sciences are equally descriptive, and to avoid the terms 'insight', 'causal connexion' and the like, which he found so obnoxious. But whatever the explanation, the result is that he talks about physics almost as though it were the Natural History of Sensations, describing 'the habits of sensations' in the way that zoologists describe the habits of zebras.

The conclusion that the sciences tell us only *how* things

happen, not *why* they do, and that all science is really an elaborate mode of description, is one that has been seized on as a lifebelt by various interested parties. Some theologians, for instance, have welcomed it as providing them with a hope of survival: if science does not aim at explaining *why* things happen, then they can continue to do so themselves, without fear of challenge from that most dangerous quarter. Their welcome has, however, been both premature and misplaced. Certainly it is no longer regarded as part of a scientist's job to say what God had in mind when He created refractive substances; so, if that is what a theologian means by 'explaining why refraction happens', a theory of refraction is not required to tell us why. But the fact of the matter is, not that physicists leave the question "What is the purpose of refraction?" to be answered by others, but that as a result of their work they no longer see this as a question which needs asking. Since the failure of Leibniz's attempts to prove that neither atoms nor a vacuum could possibly exist, 'since it would have been unreasonable of God to create them', questions about the purpose of physical phenomena have come to seem particularly fruitless—which is *not* the same as saying that scientists now regard physical phenomena as purposeless. In any case, the premise that all the sciences are alike descriptive is hardly acceptable any more. The manifest differences between the physical sciences and natural history show that this is, at best, an exaggeration, for how different are scientific explanations of the physical type from anything we could ordinarily speak of as descriptions; and how little can one think of, say, the doctrine that light travels in straight lines as 'reporting a fact' or 'describing a state-of-affairs'.

Instead of treating all sciences as equally descriptive, and explanation as metaphysically disreputable, it would be more interesting to consider how far the aims of any particular science are explanatory and how far they are descriptive. Most of the sciences which are of practical importance are, logically speaking, a mixture of natural history and physics. The nearer one is to natural history, in the agricultural sciences, for instance, the better the traditional logic-book account fits: the nearer one is to physics, the more unsatisfactory it becomes. In some subjects, such as geology and pathology, the strands are interwoven in a way which is complicated and needs examining. But the issues involved could not help being somewhat technical, and this is not the place to deal with them.

CHAPTER III

LAWS OF NATURE

FROM our study of the Principle of Rectilinear Propagation we have seen how necessary it is always to understand a physical principle in the context of its use. Looked at against this background, its force will be clear enough: divorced from all practical contexts and left to stand on its own, its meaning will be far from clear, and it will be open to all sorts of misunderstanding and misapplications. The same is true of laws of nature; and in this chapter we must try to see what the tasks of such laws are—that is to say, how they contribute to the fulfilment of the programme of the physical sciences.

3.1 *How laws of nature help one to explain phenomena*

Up to this point in our discussion, we have not come across anything that a scientist would speak of as a law of nature, for the doctrine that light travels in straight lines is not so much a 'law' as a 'principle'—the force of this distinction we shall see later. Nor have we encountered a situation in which a scientist would go in for any very elaborate experiments, so that we have yet to see the place of the laboratory in the development of the physical sciences. Nor, again, have we allowed ourselves to go beyond the kinds of phenomenon which, in the twentieth century, it does not take a scientist to explain: the study of shadow-casting hardly taxes the resources of physics. These three facts are related. It is only when we go beyond the simplest everyday phenomena to a study of more sophisticated things that resort to the laboratory becomes necessary; and it is in the form of laws of nature that the scientist ordinarily aims to express the results of the experiments he then undertakes.

We need not look far for an example to consider. When we discussed shadow-casting, we found that certain restrictions had to be placed on the circumstances in which the principle that light travels in straight lines was applied. One restriction was 'no refraction': we can use our principle confidently to argue from the height of a wall and the sun to the depth of the wall's shadow, only when there is, e.g., no glass tank of water just behind the wall, and no bonfire to produce currents of warm air and blur the shadow. It should be noticed, incidentally, that one cannot give an exhaustive list of such conditions, which does not begin with an 'e.g.' or end with the phrase 'and so on . . .', since the number of different kinds of situations in which refraction may occur is indefinitely large. Only in the absence of water, glass and the like are the techniques of geometrical optics applicable in their simplest form. So, in order to get clear about the techniques first of all we confined ourselves to everyday things, showing how the physicist's picture of optical phenomena introduces precision and system into the everyday field, and makes it possible to argue from one set of exact measurements (e.g. wall-height, 6 ft.: sun-height, 30°) to others (e.g. shadow-depth, 10 ft. 6 in.). But can we now extend the techniques of geometrical optics so as to explain also the optical phenomena we encounter in the presence of water, glass, warm air currents and the rest? This is where Snell's Law comes in.

It is worth remarking, before we go any further, that the terms in which we are here describing the investigation are not those which a scientist himself would use. What we call 'extending the range of application of the theories and techniques of geometrical optics to situations in which water, glass or other such transparent substances intervene between the lamp, or the sun, and the illuminated objects' he would call 'investigating the optical properties of transparent media'. The difference between these two ways of stating the problem arises partly from a desire for compactness, but it reflects also the differences between the attitudes which the logician— who is an onlooker—and the scientist—who is a participant— will adopt towards the symbolism of the sciences, and towards their subject-matter. Naturally enough, the scientist will always *use* his theoretical terminology in describing what he is doing. For the logician, however, the way the scientist uses his theories and symbolism is itself a part of the activity under examination: from his place in the grandstand, therefore, he will prefer to give a more cumbersome description, in which the roles of the scientist's symbolic techniques are not left unexamined, but stated explicitly.

What is Snell's Law? Let us state it first as a physicist would state it, and then go on to see how it serves to solve our problem. To use participant's language for the moment, what Snell discovered was this: that, if one measures the angles at which a ray of light is inclined as it strikes the surface of a piece of glass, water, or other transparent substance, and after passing into it, there is a simple relation between these two angles.

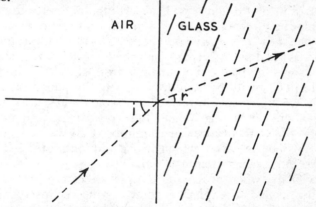

If the angle i, at which the specimen is set askew to the light striking it, is called the 'angle of incidence', and the corresponding angle r, at which the light travels after entering the glass, is called the 'angle of refraction', then Snell's Law states that

"whenever any ray of light is incident at the surface which separates two media, it is bent in such a way that the ratio of the sine of the angle of incidence to the sine of the angle of refraction is always a constant quantity for those two media."[1]

[1]The 'sine' of an angle is a simple trigonometrical function, varying from 0 for an angle of 0° to 1 for an angle of 90°, which can be found tabulated in any book of mathematical tables.

With a wide range of transparent substances, and under similar conditions, the phenomena again obey the same law, $\frac{\sin i}{\sin r}$ =const., only with a different 'constant quantity' for each substance. In the case of a few substances difficulties arise, and in these cases the refraction is said to be anomalous; but wherever the law holds in this simple manner we speak of the constant quantity for refraction out of air into the substance[1] as the 'refractive index' of the substance.

It is easy to see in outline how this law helps us. If, for instance, we find that a light-ray striking a piece of glass at an angle of incidence (i) of 60° is inclined after refraction at an angle (r) of 45°, we can at once work out what the angle of refraction will become if the angle of incidence is changed to 45°. For the ratio of the sines will, according to Snell's Law, be the same in both cases; and a little arithmetic will show that, when i is 45°, r will be about 36°. This application of Snell's Law is like inferring what the length of the shadow of a wall will be when the sun has dropped to 15°, knowing what the length of the shadow is when the sun is at 30°.

Our example is, however, still stated in participant's language, and uses terms like 'light-ray', which themselves form part of the theory we are examining. Can we, as logicians, restate the law in a way which will avoid doing this? This is what we must next attempt to do.

Previously, when producing the picture of the optical state of affairs needed to explain shadow-casting, eclipses and the like, we thought of light as propagated in straight lines (rays) from the source of light to the objects lit up, drew straight lines to represent the direction of travel of these light-rays, and remarked how they were cut off by opaque obstacles. This technique was all very well for shadow-casting, but did not explain refraction. Now we can add a new rule. When, in our picture, the straight line representing a light-ray impinges on the line representing the surface of a transparent obstacle, we

[1]Strictly speaking, this should read 'out of a vacuum into the substance', but in the case of most transparent solids and liquids the difference is trifling.

are to change its direction where it passes through the surface, and the amount of the change is to be calculated using Snell's formula. It is necessary to say 'in our picture', so that we keep in mind the fact that the lines we draw in the diagram do not necessarily stand for individual 'things' in the state of affairs represented: as we have seen, the notion of a light-ray is a theoretical ideal, which derives its meaning as much from our diagrams as from the phenomena represented, and this fact is reflected, as we shall soon discover, in the practical difficulties which limit the extent to which we can get light to travel in ever-narrower beams.

This new rule allows us to extend the inferring techniques of geometrical optics in the way we aimed to do. It also shows how the model of light as a substance travelling has to be extended to cover this new application: just as, to understand about shadows, we had to begin thinking of sunlight as travelling in straight lines from the sun to the objects it shines on, so now, to understand about refraction, we must think of the light as changing direction when it enters transparent media such as glass.

Using this new rule, we can account not only for observations made in the laboratory. We can explain also many optical phenomena which had simply to be ruled out of consideration, so long as we could employ only the more primitive techniques needed for dealing with shadow-casting. For instance, we can account for that King Charles' Head of philosophy, the stick which looks bent when its end is dipped into water.

Furthermore, when one says 'account for' the phenomenon, this does not mean coming down on one side or the other in the vacuous dispute as to whether 'in ultimate reality' the stick is bent or not: it means that, given the angle of viewing of the stick and the refractive index of water, one can actually *construct*, in a diagram of the kind given overleaf, the 'apparent position' of the stick, and so confirm that it is to be expected, light travelling as it does, that the stick will appear as it is in fact found to appear.[1]

[1]The diagram in the text has been simplified for the sake of clarity The construction shewn in fact determines only the degree of foreshortening: the exact angle at which the stick appears to be bent could be found by drawing a rather more complex diagram.

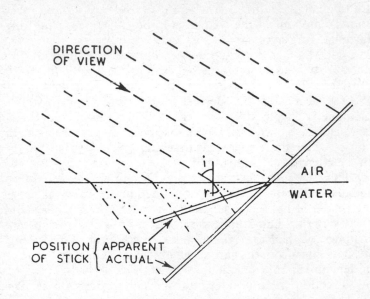

DIRECTION OF VIEW

AIR

WATER

i

r

POSITION { APPARENT
OF STICK { ACTUAL

Since this construction is as strict an application of Snell's Law as our shadow-diagram was of the principle that light travels in straight lines, one can properly say that it follows from the theory that the phenomenon must be what it is. Provided that the appropriate conditions are fulfilled, the theory can be said in these circumstances to imply the occurrence of this particular phenomenon. Arguing in accordance with the Law, that is, one can infer what will happen: unless one disputes the adequacy of the theory, therefore, one will be bound to foretell just that phenomenon in those circumstances. We can also argue in reverse. In fact, observations of a phenomenon very similar to that of the 'bent stick' are used when measuring the 'refractive index' of a substance. This fact reminds us of the virtues of the view that the physical sciences form 'deductive systems': the defects in this view we shall see shortly.

As onlookers, then, we can regard the discovery of Snell's Law as the discovery of how the optical phenomena encountered in a specifiable range of situations are to be represented, and so explained—to such-and-such a degree of precision, and with certain provisos, which we shall have to consider in a moment. This may seem to be stated vaguely, but it is inevitable that it should be: if you try to say exactly and explicitly what is involved in the discovery, with all the conditions and limitations put in, hoping to 'make an honest fact of it', you will succeed only in producing a tautology. For to cover yourself you will have either to employ at some point an omnibus phrase, like 'all relevant factors' or 'other similar situations', the nature of whose relevance or similarity cannot be independently specified, or else introduce into your provisos a circularly defined technical term like 'optically homogeneous', i.e. 'having a uniform refractive index'. But this does not mean, as some have thought, that laws of nature themselves are treated by scientists as tautologies, or as conventions: rather, it shows us one of the reasons why, in practice, the scope of a law is stated separately from the law itself—why Snell's Law, for instance, has to be supplemented by a set of statements of the form, "Snell's Law has been found to hold under normal conditions for most non-crystalline materials of uniform density." This is a distinction which will receive a more detailed examination later in the chapter.

3.2 *Establishing a law of nature* (*I*)

The discovery of Snell's Law has several features in common with the discovery we studied in the last chapter—the discovery that light travels in straight lines. To begin with, the transition from the stage at which it was not known that light travels in straight lines to the stage at which this had become known, was a double one: it involved the introduction of novel techniques for drawing inferences about shadows, eclipses and the like, and also of a novel way of thinking about the situations in which these phenomena occur—one that makes the new inferring techniques seem natural and intelligible. So here, the change which takes place when Snell's Law becomes known is also a double one: we are given a rule for extending the inferring techniques of geometrical optics to cover refractive phenomena, and the model of light as a substance in motion is deployed a little further.

Again, we found that, logically speaking, the Rectilinear Propagation Principle belonged in quite a different box from the data which are taken as establishing it; so that there can be no question of its being deductively related to these data, nor any point in looking for, or bewailing the absence of such a connexion. The transition from the everyday to the physicist's view of light involves not so much the deduction of new corollaries or the discovery of new facts as the adoption of a new approach. So now, the step from the experimental observations on which Snell's Law is based to the Law itself cannot be thought of as a matter of natural history, as a summing-up of the observations in terms with which we are already familiar. Once again, there is no question of our conclusion being either deductively related to, or a plain generalization of the observations we write down in our laboratory note-books. One might manipulate experimental apparatus for a lifetime, and accumulate all the observations one cared to, without ever spotting what form the law should take. For many centuries, indeed, scientists were with striking distance, but failed to discover it: Ptolemy, about A.D. 100, had already made many important observations on the subject but, like Roger Bacon and Kepler later on, failed narrowly to hit on the law which, in 1621, Snell at last formulated.

These things are connected with the fact that what Snell discovered was, again, the *form* of a regularity whose existence was already recognized. Ptolemy, Bacon and Kepler could not have studied refraction in the way they did unless they had been sure that there was some regularity to be discovered: indeed, it will be clear to anyone who studies the phenomena concerned that they are of a sort that cries out for an explanation. But though the existence of a regularity was clear to them, at any rate so long as one kept away from Iceland Spar and other anomalous materials, it remained to be found out what form the regularity took. This was what their experiments were designed to reveal, or rather, what they hoped to be able to spot from the results of their experiments.

To bring out the force of these points, consider how one might set about establishing Snell's Law. Let us discuss, therefore, what kind of apparatus we might assemble in order to collect suitable data. There are several important morals which we can illustrate in the course of this examination; first, about the place of experiments in the physical sciences, and secondly, about the relation between concepts of theory (such as 'light-ray') and the phenomena they are used to explain.

The question we have to ask is, in participant's language, "What happens to light-rays when they enter refracting media?"—or, to put the same thing in onlooker's language, "How are we to extend the techniques of geometrical optics to account for the optical phenomena we meet in the presence of glass, water and the like?" This is very much the sort of explicit and limited problem that we can hope to tackle experimentally. But a number of things require to be done, if we are to achieve anything:

(i) The theoretical notion of a light-ray must be given some more definite practical realization. Means are needed for producing beams of light, in the everyday sense of the phrase, which will approximate as nearly as need be to the Euclidean ideal of breadthlessness, and which will therefore be of a kind that we can accurately represent by geometrically straight lines. Until this is done, we shall have nothing that we can confidently treat as light-rays, and so nothing to study in our attempt to extend the theory and techniques of geometrical optics to the new field.

(ii) We must find out under what circumstances the phenomena of refraction will be reproducible and steady: whatever apparatus we assemble must provide us with phenomena worth investigation.

(iii) We must so arrange our apparatus that we can make measurements on it comparable with those we made when studying shadows. Only if we do so, shall we have any way of choosing how to extend the techniques of geometrical optics to the new field: otherwise the techniques will have nothing precise to explain.

These considerations are worth setting out in detail, for they can be used to illustrate an important fact. No competent scientist does pointless or unplanned experiments. There is no place in science for random observations, and only in the rarest cases have scientists made experiments whose results were of any value, without knowing very well what they were about. Before the scientist enters his laboratory at all, he must therefore have guidance about the kind of state of affairs *worth investigation*, the type of apparatus *worth assembling*, and the sort of measurements *worth making*. This guidance can come only from a careful statement of his theoretical problem, and if one looks at the conditions of the experiment he performs one will find that they are tailor-made to suit this theoretical problem.

In the present case, for instance, what is required is for the scientist to pass extra-narrow beams of light in precisely measurable directions through carefully ground prisms or lenses of unusually homogeneous glass. By arranging for the light-beams to be as narrow as possible we satisfy condition (i) —the narrower they are, the nearer they become a physical realization of the theoretical ideal of a light-ray. By demanding that our lenses or prisms be carefully ground from glass of greater than usual homogeneity we satisfy condition (ii); for only if we take some such precautions shall we find that our phenomena are sufficiently steady and reproducible to be worth studying. And by noting precisely the directions of the narrow beams of light both outside and inside the glass, we provide ourselves with observations comparable with those that we are used to dealing with in the more restricted circumstances which we have been studying up to now. Here as elsewhere, if you want to understand why a scientist is performing a particular experiment, ask how his problem came to be posed and what it was in his theory which led up to it. If you understand the theoretical problem, the reasons for the conditions of the experiment will almost certainly be clear to you: unless you understand the problem, they certainly will not.

Here again we must recognize the great differences between the physical sciences and natural history. The naturalist can afford to keep his eyes skinned from the start: it is never too soon to notice some fact of interest about the birds and animals around him. In physics, by contrast, it may easily be too soon to make any observations: until your theoretical problem has been carefully thought out, experiments will be premature. The naturalist goes about the world with an open eye and mind, prepared to notice anything of interest that may occur in his path. But the physicist does not enter his laboratory until he has some completely specific question to answer; and his apparatus will be carefully designed to extort the material he needs for an answer to this question.

Let us consider next how an experimental apparatus might be designed in order to fit our particular theoretical problem. First, there is the problem of getting light to travel in sufficiently straight and narrow beams, and in sufficiently precise directions. Normally light fans out—as the origin of the word 'ray', the Latin *radius*, reminds us, our first exemplars are the sun's rays spreading out in all directions. The difficulties one encounters when one tries to get a beam sufficiently narrow for experimental purposes are instructive, and illustrate well the nature of our theoretical concepts.

The first difficulty is a purely practical one, which raises no theoretical problems. One might begin by thinking that all one needed was a bright lamp and a single screen having a narrow slit in it:

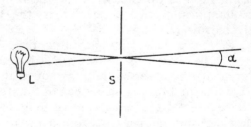

This, however, will not be satisfactory, however narrow we make the slit in the screen. Since the glowing filament of the lamp will be at least a millimetre or two across, we shall obtain not a narrow beam of light, but a fan diverging from an angle

(a) of several degrees, quite unsuitable for precise measurements. This, of course, is to be expected even on the principles of geometrical optics.

The natural next suggestion, which is the basis of all the equipment used in experiments of this kind, is to employ two screens (S_1, S_2) each with an adjustable slit in it, the slit in the first acting as the source of light for the second.

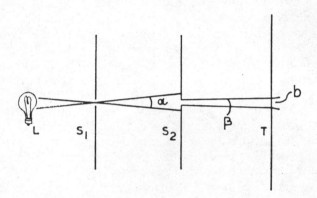

Given this set-up, there seems no reason, on the principles of geometrical optics, why we should not make the angle of divergence (β) of the resultant fan as small as we please, and so obtain as narrow a beam—as near an approximation to our theoretical light-ray—as we choose. All we need do, according to geometrical optics, is make the slits in the two screens progressively narrower.

What do we find if we set up such an apparatus? Up to a point all goes as we expect. We erect a third screen (T) as a target, and gradually make the slit in S_2 narrower and narrower; and to begin with, the breadth of the bright line (b) where our beam strikes the target decreases. But if we go on narrowing the slit, then after a certain point we get no further advantage: the only effect of doing so is to blur the line on the target, to spread it out and make it fuzzier. We are up against the phenomenon physicists speak of as diffraction.

What is the moral of this discovery? Is this the death-knell of geometrical optics—must we conclude that its principles have failed us, and must be given up? So must we abandon the hope of extending to other fields the techniques which proved so useful for explaining shadow-casting?

These reactions would be too drastic. For our discovery need only remind us that, like all techniques, the inferring techniques of geometrical optics have a limited *scope*. We can rely on them to explain a great range of optical phenomena with a high degree of accuracy, but beyond that point other methods are needed. Further, it will remind us that when we represent light by Euclidean straight lines we are setting up a theoretical ideal: it remains to be discovered from experience how far this theoretical ideal of a light-ray can be realized. Just as it is too simple to regard the discovery that light travels in straight lines as the discovery of an ordinary, but novel, matter of fact, so the term 'light-ray' as it appears in theoretical arguments must be understood as an ideal, introduced for the interpretation of the inferences of geometrical optics: it should not be thought of, so to speak, as the name of a new species of object found in a hitherto-unexplored jungle, to which we have to give a name, and whose habits it is for physicists to study.

The actual practice of scientists in such a situation as this is to recognize the existence of the limits set by diffraction, and keep clear of them in all arguments and experiments in geometrical optics. Diffraction effects will themselves be something to investigate in due course, but they are a subject for physical optics, along with other problems connected with the question "What is it that travels?" or, in physicist's language, "What is the nature of light?" : the limitations we find ourselves forced to place on the application of geometrical techniques are themselves something to be explained—though naturally something which cannot be accounted for within geometrical optics itself, but requires a richer and more refined mode of representation for its explanation. With these allowances, physicists can carry on as before. The discovery of diffraction does not prove that it is untrue that light travels in straight lines, for such a principle, as we shall see, cannot be spoken of as true or untrue in any simple sense. No more did Einstein's work prove that Newton's Laws of Motion were untrue. It accounted for some limits, which had hitherto been unexplained, to the accuracy with which Newton's mechanics can be used to calculate the motions of the planets; but it superseded Newton's mechanics only for the most refined theoretical purposes, and could only whimsically be said to prove the older laws of motion untrue.

3.3 *Theoretical ideals and the world*

It is worth while at this point considering a little more carefully the status of theoretical ideals in physics, for it is by using these ideals that the physical sciences become, as they are sometimes called, *exact* sciences.

It is easy to misconceive the nature of this exactitude, for two utterly different things have to be distinguished: the *mathematical exactitude* with which inferences are drawn in physics, and the *practical exactness* with which the conclusions of these inferences can be applied to the systems physicists study. It is the former which marks off the exact sciences from other subjects, for this exactitude is characteristic of the inferences we make in physics, genetics and the like, and is commonly absent when we turn, say, to the study of ants' eggs. The exactness of practical application, on the other hand —the degree of accuracy with which our theoretical conclusions fit the facts—is not something which marks off all the exact sciences equally, being greater in some branches than in others.

Thus in geometrical optics, using the notion of a light-ray, we can make all sorts of statements, such as Snell's Law, in exact—mathematically exact—terms. Likewise we can draw inferences, diagrammatically or trigonometrically, as exactly as we please: so far as the mathematics of the subject is concerned, we can compute the length of a wall's shadow from the heights of the wall and the sun to as many places of decimals as we choose. But all these statements and inferences will have a *physical* meaning only up to a certain point. This is not only because the sun itself has an appreciable width, so that the shadows it casts cannot in practice have more than a certain sharpness: it arises also from the fact that the arguments of physics are conducted in terms of ideals, and there is always some limit to the extent to which we have found ways either of realizing these ideals, or of recognizing bodies or systems which can be accepted as realizing them as accurately as we can measure.

Another example: if we do dynamical calculations in terms of 'rigid rods', our conclusions will again be both unique and indefinitely exact. But they will once more be about ideals: before we can draw any conclusions about the actual rods from which machines and houses are built, we must know how far the rods with which we are concerned can be treated theoretically as rigid rods, and the inferences will apply to them only as accurately as they are rigid. And what goes for rigidity goes also for other properties: there is a large family of words in the physical sciences—'rigid', 'exact', 'straight' etc.—whose members lead this kind of double life. In each case, we may contrast *either* the exactitude of mathematics with the inexactitude of experimental reports, the rigidity of the rods we argue about with the flexibility of actual rods, the perfect straightness of Euclid's lines with the imperfect straightness of any line we draw, *or* the high degree of exactness with which physical optics fits the facts with the comparative inexactness of geometrical optics, the extreme rigidity of ferro-concrete with the comparative flexibility of copper, the unusual straightness of Roman roads with the comparative windingness of most country lanes. Trouble begins in philosophy, and serious trouble at that, when we use such words as these without being clear which of the two contrasts we are intending to draw.

Furthermore, it is easy to overlook the ideal status of a term like 'light ray', and to suppose that the phrase refers simply to sunbeams and similar things. If we do this, we may be inclined to regard the doctrine of rectilinear propagation as a way of reporting such phenomena as the luminous streak which light pouring through a window makes in the air. But this will not do. For, to begin with, it is only because there are dust-motes in the air, which scatter the incoming light, that one encounters this phenomenon at all: the more visible the beam, the less completely is the light actually travelling in a straight line. In addition, the notion of a light-ray is tied to our optical explanations in a way in which that of a sunbeam is not. A child might learn to talk about sunbeams and yet have no conception of geometrical optics; but a man cannot be said to know what is meant by the term 'light-rays' if he does not understand the diagrams which we use when explaining shadow-casting. There is in fact no more direct a connexion between rays of light in the everyday sense of the phrase, such as sunbeams, and light-rays as physicists speak of them, than there is between the light which, on an August afternoon, dapples the apples and lies in great pools around the lawn and the physicist's light, which could not meaningfully be said to lie around anywhere.

This, of course, is not to deny that sunbeams are light-rays, or composed of light-rays. Certainly we shall often be able to apply to sunbeams the inferences that we draw in terms of light-rays: we did this without hesitation in calculating the depth of the wall's shadow. It is, rather, to mark the distinction *in logic* between words like 'sunbeam' and phrases like 'light-ray', i.e. to draw a distinction of logical type, like that between the person and name of Winston Churchill and the title and office of Prime Minister; and this can be done regardless of whether or not *in fact* Winston Churchill at present holds the office of, and so is describable as Prime Minister.

Similar distinctions are important when one examines the use which is made in geometry and physics of the terms 'point', 'particle', etc. Old-fashioned text-books tend to start off with mystifying definitions of these terms: Euclid's own definition, "A point is that which has no part", is a good example. After a perfunctory discussion of these, the author clears his throat, begins a new chapter and gets going with some concrete examples: the definitions are mercifully forgotten. And this is as it should be. Definitions of these terms are not called for, and the more self-conscious authors of text-books are at last ceasing even to go through the motions of defining them. For the questions to be asked about points, particles and the rest are not "What *is* a point?", "What *is* a particle?" etc.:

they are "What *can be regarded* for physical purposes as a point, particle, etc.?" Or rather, since we soon find out that under some circumstances or other almost any region of space can be treated as a point, and almost any body—even the sun— as a particle, the sort of question to be asked is, "Under what circumstances can the sun, say, be regarded as a particle?"; or, what comes to the same thing, "Under what circumstances can the inferences we make in terms of particles in our dynamical calculations be applied to the sun, and its dimensions be neglected?" A particle in dynamics is not 'an indefinitely small material object': if one insists on a definition, it is 'any material object whose dimensions can, for the purposes of the present calculation, be neglected'.[1]

This brings us back to the notions of exactness and exactitude. For in practice we shall always have to ask, not "Is an aeroplane a particle, or a sunbeam a ray of light?", but "Under what circumstances and with what degree of exactness, i.e. accuracy, can one treat an aeroplane as a particle for dynamical purposes, or a sunbeam as a ray of light for optical ones?" The inferences of physical theory remain in every case exact: it is the accuracy with which the conclusions are applied that varies.

3.4 *Establishing a law of nature* (*II*)

So much for the first of our problems, that of realizing our theoretical ideal of a light-ray. Let us suppose, then, that we have assembled a pair of screens with narrow slits in them, a bright lamp, and a target screen, and that the slits are set to provide a beam which is as narrow as is practicable, bearing in mind the limits that we have been discussing. Now we have produced some light rays, or near enough, what about our refracting medium?

At this point we encounter the second of the practical problems facing us: how to ensure that we have steady and reproducible phenomena to study. If we set up the same apparatus two days running and go, to the best of our know-

---

[1] These remarks do not apply as they stand to the 'fundamental particles' of atomic theory.

ledge and belief, through identical steps each time, and the optical phenomena we observe on the two days are markedly different, we are clearly in no position to make any worthwhile observations: still more so, if we set up the apparatus and the phenomena fluctuate under our very eyes.

Any experimental set-up in a laboratory is inevitably a highly artificial one. When it comes to studying refraction, say, especially with such a specific end in view, one cannot hope to find suitable specimens for one's experiments simply lying around. Notice, incidentally, the contrast with natural history: the naturalist must take his frogs as he finds them. Nor could one be confident that one's apparatus was going to satisfy all the required conditions if it consisted merely of a collection of *objets trouvés*. Such more or less transparent objects as one might pick up would be of largely unknown physical and chemical composition and of unsuitable shapes, whereas we require to study only objects whose characteristics we know, and whose shape in particular we can precisely control: hence the demand for accurately ground prisms. Further, if we used any glass a manufacturer happened to supply, we might still find that the rays in the glass tended to waggle: we must therefore get the manufacturers to supply specially homogeneous, so-called 'optical' glass, carefully mixed and slowly cooled for consistency. It is worth noticing, by the way, that this recipe involves a hypothesis that the optical properties of a material depend on the constancy of its density and on its degree of homogeneity: how far this is the case is something which requires independent investigation. Again, if we are careless about temperature variations, we shall find our results varying: certain precautions will have to be taken—keeping Bunsen burners away from the apparatus and shading it from the sun— if our experiments are to be fruitful.

These, in the present example, are the most important precautions. If others were needed in order to control relevant factors and ensure steady and reproducible results, they could no doubt be taken. But just what steps will be needed, just what factors are relevant to any question and therefore have to be controlled in an experiment, is something which will have to be found out: there can be no general recipe. In this respect, the demand for homogeneous glass or the avoidance of temperature variations is on a different footing from the precautions we take in order to obtain narrow beams of light with precisely measurable directions: these last steps are essential, not for practical reasons, but for theoretical ones, in order that the apparatus shall be capable of helping us to the solution of our theoretical problem.

The list of precautions may, in some experiments, be fairly long, but it will always be finite and definite. If an experiment gives an unexpected result, the conclusion that some relevant factor must have been overlooked is normally acceptable only when a possible factor can be suggested and investigated: perhaps the test-tube was not clean. In a well-planned experiment, this can be checked for all the factors which there is any reason to consider as relevant. So, though in fact 'escape-clauses' of this kind may sometimes be invoked, one cannot do so arbitrarily, merely in order to preserve a particular theory from discredit. We are not forced, accordingly, to speak of physical theories as conventional: the discovery that for some phenomenon to occur as it normally does, some factor which is constant under normal experimental conditions, magnetic field gradient, say, has to remain constant, may be a major discovery. The Zeeman and Stark effects might be quoted as examples of this sort of discovery: it would not at first occur to one that the kind of light radiated by a body depended on the strength of the magnetic and electric fields to which it was exposed.

The apparatus with which we shall study refraction can be thought of, then, as consisting of three things: a source of light arranged to emit as narrow a beam as is practicable, a parallel-sided specimen of the material being studied, carefully made and mounted so that the direction at which the beam strikes it can be accurately measured, and a target screen or other device for observing how much the specimen deflects the light which passes through it.

Two questions now need to be asked: what sort of observations shall we make with this apparatus, and how will they be connected with the conclusion we are using them to establish, viz. Snell's Law?

There are various different sorts of observations we might make: it is enough to consider a typical one. Supposing, therefore, that we have arranged the specimen to swivel through any angle we please, let us set it successively at angles to the beam of 0°, 5°, 10°, 15° . . . and so on.

As we turn it progressively more and more askew, the bright line on the target will be shifted more and more from its first position. Let us make a note of the amount of the deflection (x) corresponding to each angle (i) at which we set the specimen: it will be a matter of simple geometry to compute from the amount of the deflection the angle of refraction (r) of the light-beam within the specimen. The results can then be tabulated in three columns: 'setting of specimen, i', 'reading of deflection, x', and 'corresponding angle of refraction, r'. The figures we write down will in each case be subject to a 'probable error', to allow for inaccuracies in the measuring technique, the grinding of the specimen and so on. It is always enough that the predictions of theory should largely fall within the region marked off by the probable error: one does not need to insist that every reading made should tally exactly with the theory.

What, now, is the connexion between the figures we have tabulated in our note-book and the law we are using them to establish? Looking at the observations and the law from a logician's point of view, what shall we say is the relation between them? There is certainly no deductive connexion either way between Snell's Law and the set of statements,

"When the specimen was set at 5°, the deflection of the beam was 2 mms." Nor is the law to be thought of as a simple generalization of the experimental results, despite the words 'whenever' and 'always' appearing in the formulation of it quoted earlier on. These words are used misleadingly, for the law is no more a universal generalization than the Rectilinear Propagation Principle turned out to be: presented with any of the situations and substances to which it ceases to be applicable, a physicist will bring into the open his unstated qualifying clauses—'anomalous refraction apart', 'the specimen being homogeneous' and the rest. Leaving aside these clauses, which are concerned with the application of the law, we are left with the statement of the form of a regularity—that the sines of two angles are in a constant ratio—and the value of the experiment is to show just how accurately this form of regularity fits the observed phenomena. "When the sun was at 30°, the shadow of a 6 ft. high wall was 10 ft. 6 in. deep . . . : *ergo*, light travels in straight lines," "When the specimen was set at 5°, the deflection being 2 mms., the angle of refraction was 3° . . . : *ergo* the ratio of the sines of the angles of incidence and refraction is constant": though these two steps are by no means identical in type, they are at any rate alike in conforming tidily to neither of the standard logic-book patterns of argument.

3.5 *The structure of theories: laws, hypotheses and principles*

In the last chapter, we remarked briefly on the special logical character of the nature-statements we meet in physical theory, and on the systematic character of scientific, as opposed to everyday language. These are things which can be made more intelligible with the help of the examples we have looked at in the present chapter. By noticing how the different types of statement we encounter in the theory of refraction are logically related, we can see what people have in mind when they speak of such theories as forming hierarchical or deductive systems.

Notice for a start, then, how the way in which physicists handle their theoretical statements marks them off from the familiar statements of everyday life, and from those of the naturalist. First, since Snell's Law is stated in terms of 'light-rays', it can be given a physical meaning only in circumstances in which the term 'light-ray' is intelligible, i.e. within the scope of geometrical optics. Where the optical phenomena are not such as are explicable in terms of geometrical optics, Snell's Law ceases even to be interpretable. Secondly, it is the practice in the physical sciences to leave the application of a law to be shown or stated separately: indeed, this itself is rather a misleading thing to say, for that this should be done is not so much a question of practice as the distinguishing mark of a law. The statement, "Most transparent substances of uniform density, excluding only certain crystalline materials, such as Iceland Spar, have been found to refract light in such-and-such a manner" is not what we call 'Snell's Law'. This statement is a simple report of past fact, and its job is to tell us about the circumstances in which Snell's Law has been found to hold. To every law there corresponds a set of statements of the form "X's law has been found to hold, or not to hold, for such-and-such systems under such-and-such circumstances." Further, in order to discover how far this range of substances and circumstances, i.e. the 'scope' of the law, can be extended, a great deal of routine research is undertaken, research which can in no way be said to call in question the truth, or the acceptability, of the law itself.

This feature is one which is not shared in everyday speech even by those statements which Ryle calls 'law-like statements', such as "Glass is brittle". When a manufacturer produces a new type of glass of exceptional toughness and resilience, we say, "All glass except Tompkinson's Tuffglaze is brittle", not "'Glass is brittle' holds for all glass except Tompkinson's Tuffglaze". This invention certainly affects the truth of our initial statement: after it, the law-like statement is said to be 'true on the whole—but not true of Tompkinson's Tuffglaze', whereas before the invention it had been 'true universally'.

Laws of nature, however, are different: to them the words 'true', 'probable' and the like seem to have no application.[1] To begin with, perhaps, we may suppose that light-rays are

[1]At any rate if we ask, "Is this law true?": on the other hand we *can* ask the question, "Is this the true (form of the) law?"

always bent by transparent media in the way they are by the glass specimen in our apparatus. We may, therefore, adopt Snell's formula tentatively, hypothetically, as a guide to further experiments, to see whether the phenomena always happen so. On this level, we might ask "Is Snell's *hypothesis* true or false?", meaning "Have any limitations been found to the application of his formula?" But very soon—indeed, as soon as its fruitfulness has been established—the formula in our hypothesis comes to be treated as a *law*, i.e. as something of which we ask not "Is it true?" but "When does it hold?" When this happens, it becomes part of the framework of optical theory, and is treated as a standard. Departures from the law and limitations on its scope, such as double refraction and aniso-tropic refraction, come to be spoken of as anomalies and thought of as things in need of explanation in a way in which ordinary refraction is not; and at the same time the statement of the law comes to be separated from statements about the scope and application of the law.

In this last respect, laws of nature resemble other kinds of laws, rules and regulations. These are not themselves true or false, though statements about their range of application can be. Suppose there is a College rule against walking on the grass: one can ask how widely it applies—whether there is any class of people, such as Fellows, who are not bound by it. Accord-ingly, statements can be made about the rule which can be true or false. If it is said that, despite the rule, Fellows are allowed on the grass, one may reasonably ask, "Is that true?" But one will not ask "Is the rule true?", nor will physicists ask this of a law of nature.

This must not be misunderstood. Suppose one says that laws of nature are not true, false, or probable; that these terms are indeed not even applicable to them; and that scientists are accordingly not interested in the question of the 'truth' of laws of nature—all of which might fairly be said: one does not thereby deny the obvious, namely, that scientists seek for the truth. One points out, rather, that the abstract noun 'truth' is wider in its application than the adjective 'true', that different types of statements need to be logically assessed in different terms, and that not every class of statement in which a scientist deals need be such as can be spoken of as 'true'/'false'/'probable'. This, of all things, is most often overlooked in the logical discussion of the physical sciences: it is therefore essential to insist on it. Saying a law holds universally is not the same as saying that it is true always and not only on certain conditions. The logical opposition 'holds'/'does not hold' is as fundamental as the opposition 'true'/'untrue', and cannot be resolved into it.

Further, laws of nature are used to introduce new terms into the language of physics—the term 'refractive index', for instance—and such things as refractive index become in their turn subjects for research. How, we may ask, does the refractive index of a substance depend on its temperature? How, to use onlooker's language, would we have to alter our ray-diagrams in order to account for the way in which the optical phenomena are affected by heating up the specimen, or for such things as the shimmering in the air over a bonfire? Notice one thing in particular: that questions about refractive index will have a *meaning* only in so far as Snell's Law holds, so that in talking about refractive index we have to take the applicability of Snell's Law for granted—the law is an essential part of the theoretical background against which alone the notion of refractive index can be discussed. This is something we find generally in physical theory. Theoretical physics is *stratified*: statements at one level have a meaning only within the scope of these in the level below.

This fact must be borne in mind when we consider the distinction between the hypothetical and established parts of physics, for this is a distinction which has been widely misconceived. It has been said by some philosophers, for example, that *all* empirical statements are hypotheses, which can, strictly speaking, never be called more than 'highly probable': in support of this view they have pointed out that we could always, by a sufficient stretch of imagination, "conceive the possibility of experiences which would compel us to revise them". Now it is important to recognize what violence this sort of argument does to the terms 'hypothesis' and 'hypothetical'. For although all the statements we meet in science are such that one can conceive of their being reconsidered in the light of experience (i.e. empirical), only some of them can, in the present sense, be called 'hypothetical'. We are now in a position to see why this is.

One can distinguish, in any science, between the problems which are currently under discussion, and those earlier problems whose solutions have to be taken for granted if we are even to state our current problems. One cannot at the same time question the adequacy of Snell's Law *and* go on talking about refractive index. But the fact that, at any particular stage, many of the propositions are taken without question does not make the exact sciences any the less empirical: it merely reflects their logical stratification. Certainly, every statement in a science should conceivably be *capable* of being called in question, and of being shown empirically to be unjustified; for only so can the science be saved from dogmatism. But it is equally important that in any particular investigation, many of these propositions should not actually be called in question, for by questioning some we deprive others of their very meaning. It is in this sense that the propositions of an exact science form a hierarchy, and are built one upon another; and just as a bricklayer is only called upon at a given moment to determine the positions of the bricks in a single course— which in their turn will become the foundation for the next course—so the scientist is only called upon at any one time to investigate the acceptability of statements at one level. Now and then there may have to be second thoughts about matters which had been thought to be settled, but when this happens, and the lower courses have to be altered, the superstructure has to be knocked down, too, and a batch of concepts in terms of which the scientist's working problems used to be stated— 'phlogiston' and the like—will be swept into the pages of the history books. But for the time being it is only the top course of bricks, the matters which are actively in question, which the scientist has to deal with. From this we can see why the discovery of phenomena which can be treated as standards and of laws which can, to use a phrase of Wittgenstein's, be put in the archives, is an essential step in building up a fruitful body of theory.

The terms 'established' and 'hypothetical', as used in science, need to be understood in terms of the distinction between the parts of a science that are actually being called in question, and those which we must take for granted in order to state our working problems. It is the statements that figure in the latter parts which are spoken of as established. Even a few of these may eventually have to be reconsidered, but there is no need—nor are we in a position—to anticipate the day when this will happen. These parts provide the background against which current problems are considered, and give a meaning to the terminology in which they are stated. The statements which we meet in them will be of two kinds: on the one hand, laws of nature, and on the other, statements about how far and in what circumstances these laws have been found to hold. Neither of these classes of statement need, or can, be spoken of as 'only highly probable': the experimental reports are not unlimited generalizations, but simple statements of past fact, while the laws of nature are not the sorts of thing we can speak of as true, false or probable at all. Yet both can reasonably be called empirical.

Contrasted with the established parts of a science, there are those problems the solutions of which are not yet clear, and about which we can at the moment say only tentative, hypothetical things. These questions are indeed open, undecided, matters for 'hypotheses'. But the statements in these hypothetical parts of a science depend for their very *meaning* upon the acceptability of the lower levels of theoretical statement; so we are debarred from speaking of the established propositions as being hypothetical also, unless and until they themselves become once again the subjects of active doubt. It could be correct to speak of *all* empirical propositions as hypotheses only in a language which was entirely devoid of logical stratification—the language of a people without any science. This stratification is a feature of the theoretical sciences in particular, as is borne out once more by the contrast with natural history. We should not so much mind saying that the generalizations of natural history can be at most highly probable: next year a pig *might* fly.

The distinction between laws and hypotheses is therefore a logical matter, involving far more than the degree of confidence with which we are prepared to advance them, or the number of confirming instances we have observed. But what about the distinction between laws and principles? Why is the Rectilinear Propagation of Light called a 'principle' and Snell's Law a 'law'?

This distinction turns upon something we noticed earlier: namely, the role of the principle as the keystone of geometrical optics. One can quite well imagine a geometrical optics in which the law of refraction was different. The adoption of a different law in place of Snell's Law would, of course, mean considerable changes—our present notion of refractive index would be one casualty. But geometrical optics could still exist as a subject, and designers of optical instruments, having learnt the new rule for tracing the passage of rays through their assemblies of lenses, would soon accommodate themselves to the change. By comparison, the principle that light travels in straight lines seems to be almost indefeasible: certainly it is hard to imagine physicists abandoning completely the idea of light as something travelling in straight lines, for to give up this principle would involve abandoning geometrical optics as we know it. If we question the principle of rectilinear propagation, the whole subject is at stake: that is why the principle is not open to falsification in any straightforward way.

It is not that, for physicists, the principle ceases to be empirical and becomes tautologous or conventionally true. They might, in circumstances sufficiently unlike the present ones, decide to give it up entirely, but they would do so only if they were ready to *write off* geometrical optics as a whole. What the circumstances would have to be, in order for physicists to decide that the methods of geometrical optics were no longer any use, is something that is open to discussion, but this would clearly require changes in the world far more drastic than those which are needed to falsify any naïve interpretation of "Light travels in straight lines", e.g. as an empirical generalization.

It is the middle-level propositions in the hierarchy of physics which alone are called 'laws', and they alone have an ambivalent logical status. Such a proposition as Snell's Law begins as an element in a hypothesis within geometrical optics, something which cannot be explained without talking about light-rays; but later it becomes an established part of the theoretical background, while the foreground is occupied by other propositions which have a meaning only where the law holds. Since its place is *within* geometrical optics, to change the form of the law is not to raze a whole subject to the ground. There is, by contrast, no body of theory against which the proposition that light travels in straight lines can be set. It is as though this principle enshrined in itself the geometrical mode of representation; and it can be discussed, accepted or rejected on one level only.

One last point about the stratification of physical theory: this is sometimes presented in a misleading way. It is suggested that the relation between statements at one level and those at the next is a deductive one, and the resulting hierarchy is accordingly spoken of as a 'deductive system'. One is given the idea that physical theories form a logical pyramid, with the straightforward reports on our experimental observations at ground level, and above them layer upon layer of progressively more general generalizations. One can illustrate the sort of thing envisaged by supposing it to be discovered that rodents consume milk-products: this would be two layers up, since from it we can deduce both "Mice eat cheese" and "Rats drink milk", and from these again we can deduce, e.g., that a mouse which we now have under observation will eat the cheese we are about to offer it.

As here presented, the picture is open to several objections. To begin with, the role of deduction in physics is not to take us from the more abstract levels of theory to the more concrete: as we have seen, these cannot, as Mach supposed, be thought of as deductively related one to another. Where we make strict, rule-guided inferences in physics is in working out, for instance, where a planet will be next week from a knowledge of its present position, velocity and so on: this inference is not deduced from the laws of motion, but drawn in accordance with them, that is, as an application of them. Nor are statements in terms of 'refractive index' deduced from Snell's Law. There is a logical connexion between them, certainly; but this is because the term 'refractive index' is introduced by reference to Snell's Law, and not because the two classes of sentences can be deduced from one another. It is the *terms* appearing in the statements at one level, not the statements themselves, which are logically linked to the statements in the level below.

One thing in particular would be especially mysterious on the deductive system account; namely, the status of the most abstract statements of all. For, if things were as suggested, each of these would be the assertion of a tremendous coincidence: if it were a coincidence that not only did mice eat cheese but also rats drank milk, so that we could daringly generalize that rodents consume milk-products, how much more of a coincidence must one regard, say, Einstein's most abstract physical principles. Further, like all such coincidences the most abstract statements of all would be particularly open to sudden upset; for surely some obscure South American rodent might turn out to be entirely herbivorous, and, if so, how much less likely still that no exception would ever be found to Einstein's theories. But this is a caricature. It is clear from a study of Einstein's work that he is concerned, not with daringly wide generalizations from experiment, but rather with conceptual matters: such equations as Einstein's certainly do not have the status the pyramid-model allots to them. Indeed, it is no accident that one has to resort to habit-statements about rats, mice and the like, in order to illustrate the point of the pyramid-model; for, however it might do as a picture of natural history, it misrepresents the logical structure of theoretical physics.

3.6 *Different kinds of laws and principles*

In this chapter, as in the last, we must ask how many of the things we have noticed about the example under detailed examination apply more generally. How far, then, can we regard Snell's Law as a typical law of nature?

Many of the things we have said about it would not be true of all laws equally, for there is a wide range of things which are spoken of in physics as laws of nature. At one extreme, one finds statements of the sort which are sometimes called 'phenomenological laws'. These involve no theoretical terms at all, not even such elementary ones as 'light-ray': a good instance is Boyle's Law, which states that the pressure and volume of a gas vary inversely at a given temperature. At the other extreme, one has such laws, or sets of laws, as Newton's three Laws of Motion, or Maxwell's Laws or Principles of Electromagnetism: these are not used directly to express the form of a regularity found in phenomena, as Boyle's Law is, but are rather like the axioms of a calculus, which are accepted so long as applications of them are found in practice to fit the facts. It will be the test of such comparatively abstract laws, not so much that they account directly for the observed phenomena, as that they provide a framework into which can be fitted the phenomenological laws which in their turn account for the phenomena. Snell's Law is of an intermediate kind, though one which is nearer to Boyle's Law than to Newton's three Laws; while the most abstract laws—like Maxwell's Principles of Electromagnetism and the Principles of Thermodynamics—come in time to have a position in their subjects almost like that of the Rectilinear Propagation Principle in geometrical optics, and are perhaps spoken of more naturally as principles thanas laws of nature.

Since the parts which different laws of nature play are so very different, one cannot expect them to have many features in common. But one such feature they do have; and it is one which, in the case of Snell's Law, proved of the first importance. They do not tell us anything about phenomena, if taken by themselves, but rather express the form of a regularity whose scope is stated elsewhere; and accordingly, they are the sorts of statements about which it is appropriate to ask, not "Is this true or not?" but rather "To what systems can this be applied?" or "Under what circumstances does this hold?"

Boyle's Law, of all laws of nature, looks most as though one could ask of it, "Is this true or not?"; yet even it would nowadays be treated in a way which rules this out. We know very well that in some comparatively unusual circumstances gases can be shown to behave in ways markedly at variance with Boyle's Law; and that at all temperatures their behaviour deviates from it to a minute but measurable extent. For theoretical reasons, as well as reasons of convenience, however, it is preferable not to regard this as a reason for scrapping Boyle's Law, but to keep the law in circulation as a first, more-or-less approximate expression of the way gases behave. The extent to which, in different circumstances, the observed behaviour of gases conforms to or deviates from it is then recorded separately; and accordingly the question whether the law is true or not no longer arises.

There are indeed certain laws in physics that one might take, at first sight, for exceptions to the rule that laws of nature are not 'true or untrue' but rather 'hold or do not hold'; for instance, Kepler's three Laws of planetary motion. These laws tell us, among other things, that the planets move round the sun in ellipses, and they are unquestionably statements about which one can ask, "Is this true or not?"—if they correctly represent the orbits of the planets, they are true: if not, they are untrue. But along with this difference go others, which show the force of our rule. For Kepler's Laws set out to tell us, not about planets in general, but about *the* planets, viz. Mercury, Venus, etc.; they summarize the observed behaviour of all members of this class, and do not set out to explain it in terms of the nature of things; they are thus even more completely phenomenological than Boyle's Law; and correspondingly no physicist would ever speak of them as laws *of nature*. One could, no doubt, formulate three nature-statements, each of which corresponded to one of Kepler's three Laws; but in order to qualify as laws of nature these would have to be expressed, not in terms of 'the planets', but in terms of 'bodies moving under the influence of gravitation alone'. Such laws would be the means, *inter alia*, of explaining Kepler's observational laws; but to identify them with Kepler's Laws would be a mistake, since it would mean overlooking one logically crucial step—that of recognizing that 'the planets', viz. Mercury, Venus etc., qualify for theoretical purposes as bodies moving under gravitational attraction alone. As Wittgenstein points out in the *Tractatus*, "The description of the world by mechanics is always quite general. There is, for example, never any mention of *particular* bodies in it, but always only of *some bodies or other*."

The status of such sets of laws as Newton's three 'Axioms, or Laws of Motion' is something which philosophers have found perennially puzzling. Those students who take the ordinary scientific training in dynamics find this question passed over in text-books with a few embarrassed and inconsistent remarks. Experimental physicists like to talk as though the laws were purely phenomenological; but this suggestion is discredited by the discovery that three technical terms, 'mass', 'force' and 'momentum', are introduced into the subject along with the three statements. After this, it is not surprising if logicians who come to dynamics from a study of ordinary discourse feel that the whole proceedings are tautological, and the argument that the laws are thereby shown to be conventional becomes attractive.

Each of these doctrines is in its way equally misleading, for the true status of the laws of motion can be seen clearly only if one examines in detail how they in fact enter into dynamical explanations.[1] When this is done, one finds that both the everyday models with which one is tempted to compare them are unsatisfactory. Newton's Laws of Motion are not generalizations of the 'Rabbits are herbivorous' type; but they are not for this reason any the more tautological (cf. 'Rabbits are animals'); and this is because they do not set out by themselves to tell us anything about the actual motions of particular bodies, but rather provide a form of description to use in accounting for these motions. The heart of the matter is put forcibly, and almost to the point of paradox, in a celebrated passage of Wittgenstein's: "The fact that it can be

---

[1]Axiomatic theories really need a chapter to themselves: here there is room only for the briefest of remarks about them.

described by Newtonian mechanics tells us nothing about the world; but *this* tells us something, namely, that it can be described in that particular way in which as a matter of fact it is described." But we must notice that it is no denigration of a system of mechanics to say that, by itself, it tells us nothing about the world. This is not to say that it fails to do what it sets out to do: it is to recognize its proper ambitions. As we saw earlier, a description of the techniques of geometrical optics by itself tells us nothing about shadows; for this, we must find out also how far and under what circumstances these techniques can be employed. So also, laws of nature express only the forms of regularities: the burden of our experimental observations is borne, not by them, but by statements about when the laws of optics hold, or how the laws of motion are to be used to represent the actual motions of planets, projectiles, leaves, ships and waves. There is, so to speak, a division of labour in physics, between laws themselves and statements about the ways in which, and the circumstances in which laws are to be applied. It is by recognizing the nature of this division that one comes to see how physicists steer their way between the Scylla of fallible generalization and the Charybdis of empty tautology.

If we are asked what the job of Newton's laws is, we may not know at first whether to say that they describe the way things move, define such terms as 'force', 'mass' and 'momentum', or tell us about the mode of measurement of force and the rest. But there are very good reasons for this uncertainty. The laws themselves do not do anything: it is we who do things with them, and there are several different kinds of things we can do with their help. In consequence, there is no need for us to be puzzled by the question whether Newton's Laws are descriptions, definitions, or assertions about methods of measurement: rather, it is up to us to see how in some applications physicists use them to describe, say, the way a shell moves, in others to define some such quantity as electromotive force, and in others again to devise a mode of measurement of, say, the mass of a new type of fundamental particle. It is not that the laws have an ambiguous or hazy status: it is that physicists are versatile in the applications to which they put the laws.

3.7 *Locke and Hume: Are laws of nature necessary or contingent?*

In the light of this discussion of laws of nature, it will be worth while examining the views philosophers have put forward about them, to see how far these views truly reflect the uses to which laws of nature are put in scientific practice, and how far the disagreements that have arisen are a sign rather of confusion or cross-purposes. But before we come to this, it is important to do one thing: namely, to distinguish between four different classes of sentence that one meets in books of physics. When scientists use the word 'law', they do not always trouble to show which class of statements they are referring to, though when they do their usage is the one we have adopted: only rarely, in fact, is there any strong reason for them to draw these distinctions explicitly. As logicians, however, we cannot afford not to distinguish between the various classes, since they have markedly different logical characteristics; and in the past philosophers have sometimes been less careful to do so than they might have been.

The four classes of statement are the following:

(i) abstract, formal statements of a law or principle—e.g. Snell's Law, in the form quoted above;

(ii) historical reports about the discovered scope of a law or principle—e.g. the statement that Snell's Law has been found to apply to most non-crystalline substances at normal temperatures;

(iii) applications of a law or principle to particular cases— e.g. the statement that, in a particular prism now under examination, the directions of the incident and refracted beams vary in accordance with Snell's Law; or the statement that the sunlight getting over a certain wall is travelling to the ground behind the wall in a straight line;

(iv) conclusions of inferences drawn in accordance with a law or principle—e.g. the conclusion that, the angle of incidence and refractive index being what they are, the angle of refraction must be 36°; or the conclusion that, with the sun at 30°, the shadow of a 6 ft. high wall must be 10 ft. 6 in. deep.

The main types of theory philosophers have put forward about the logical character of laws of nature are also four in number. This is no coincidence, for one finds exponents of the four views citing, as support for their accounts, facts about the appropriate one of our four types of statement. Accordingly, these views may not really be the irreconcilable rivals they have seemed. Perhaps their appearance of opposition reflects rather a preoccupation with different aspects of laws of nature. How far this is so, we must now consider.

On the one hand, then, one finds it suggested by Locke, and more recently by Kneale, that laws of nature are principles of natural necessitation, comparable with statements like "Nothing can be both red and green all over" except in one respect—that where the necessity of the latter is something we can 'see', the necessity of laws of nature is not immediately visible, i.e. obvious, but is rather forced on us as a result of our experiments. The metaphors 'transparent' and 'opaque to the intellect' have been used by Kneale to mark the difference between them. This view has been found objectionable by such philosophers as Hume and Mach: they have felt that nothing which a scientist can properly be said to discover could be, in the logical sense, necessary, and they have accordingly preferred to advance the theory that laws of nature are statements of constant conjunction, which tell us that such-and-such sets of characteristics have always been found to go together. A third view, designed to circumvent the traditional problems about induction, is that which Kneale attributes to Whitehead: according to this, laws of nature should be regarded as conjectures about uniformities holding over limited regions of space, for limited periods of time, i.e. not as universal generalizations, but rather as generalizations supposed to be true throughout a vast but not infinite region and period of time surrounding our own—what may be called a 'cosmic epoch'. Finally, Moritz Schlick and F. P. Ramsey have argued that laws of nature are not "propositions which are true or false, but rather set forth instructions for the formation of such propositions . . . [being] directions, rules of behaviour, for the investigator to find his way about in reality."

We must now notice, in turn, how each of these theories reflects some aspect of the uses to which principles and laws of nature are put in the physical sciences. Let us begin by looking at the Lockean theory, that laws of nature are principles of necessitation. To recognize the force of this view, recall the way in which physicists use such words as 'must', 'necessarily' and so on, especially in the conclusions of their arguments— cf. class (iv) above. In our first sample explanation, for instance, we saw how a scientist will say that, the height of the sun being 30°, and that of a wall being 6 ft., the shadow of the wall is necessarily 10½ ft., and indeed that it follows from, or in accordance with, the Principle of the Rectilinear Propagation of Light that it must have just that depth and none other. It is clear from this that, in some sense or other, physicists do treat their laws and principles as telling us, or enabling us to discover how things *necessarily* are, and what in given circumstances *must* happen; and the phrase 'principles of necessitation' is presumably intended to reflect just this sort of fact about laws of nature.

What needs to be made clear, however, is that the sense in which one can speak of laws of nature as telling us how things 'must' happen is not one that need be obnoxious to Mach and Hume. So let us ask again: when the physicist says that it follows from his principle that the shadow must have just such a depth, what kind of inference is this, and what sort of necessity? How can it be said to follow from any experimentally established principle that the depth of the shadow *must* be what it is?

To answer these questions correctly one must distinguish between two pairs of things: first, between establishing a theory and applying an established theory; and again, between recognizing a situation as one in which a particular theory can be employed, and employing the theory in that situation on the assumption that it has been correctly identified. It is part of the art of the sciences, which has to be picked up in the course of the scientist's training, to recognize exactly the situations in which any particular theory or principle can be appealed to, and when it will cease to hold. Although a scientist can often say what it is about one situation or another which makes a particular theory applicable or inapplicable, there is always a certain amount of room for the exercise of individual judgement; and this makes it nearly as difficult to give rules for deciding when a theory must be modified or abandoned as it is to give rules for discovering fertile new theories. But provided that the scientist has correctly identified the situation with which he is faced for what it is, and therefore knows what principles and laws he can appeal to, it is the very business of the theory to tell him what *must* happen, i.e. what he must expect to happen, in such circumstances. If this is a field of study which has been brought within the ambit of the exact sciences at all, his theory will provide him, among other things, with an inferring technique—that is, with a way of arguing from, e.g., the height of a wall and the angle of elevation of the sun to the depth of a shadow. The actual technique of inference-drawing may be a geometrical one, in which one draws inferences by drawing lines, or it may be a more complicated, mathematical one. But in either case it is essential, if the theory is to be acceptable, that it shall license one to pass in one's arguments from the conditions in which the particular phenomenon takes place to the characteristics of the phenomenon which are to be predicted or explained.

Now there is nothing that need worry Hume in the use which, as a result, the physicist makes of words like 'must' and 'necessarily'. For when he says, "In those circumstances the shadow *must* be ten and a half feet deep", he does so always with the tacit qualification, "If all the conditions are indeed fulfilled for the application of this principle"; the depth of the shadow is therefore not a necessary fact, but a necessary consequence of applying the principle as it is meant to be applied. And when we say that it follows from the principle that, in such circumstances, the shadow *must* have that particular depth, the principle finds its application, not as a major premise in a syllogistic argument from generalization to particular instance, but as the 'inference-ticket', to use a phrase of Ryle's, which entitles us to argue from the circumstances of the phenomenon to its characteristics. In the circumstances of our example, it has been found that shadow-casting and the like are explicable, representable or predictable in a way which makes use of certain geometrical and trigonometrical relations: arguing in accordance with the rules which express these relations, one *must* in these particular circumstances expect the shadow to be just the depth it is. It is because, and only because a physical theory involves techniques of inference-drawing that a 'must' enters in. Once we have been taught such a technique, a correctly performed computation of the depth of the shadow *must* lead to the result it does, and any computation which fails to lead to this result *must* be faulty.

Hume and Mach are, nevertheless, justified in insisting on this: that the possibility of explaining particular phenomena in a particular way is something which *has to be found out*. One could not say that the techniques of geometrical optics must be applicable in the ways in which they have been found to be applicable, except in so far as this fact is, in its turn, explicable by reference to a wider theory. One might, perhaps, appeal to the wave-theory of light in order to show that ray-diagrams must be applicable just when they are found to be; but this simply shifts the burden. The important thing is not to confuse the questions, what theory *has been found* reliable in a given field, and what phenomena, according to this theory, *must* occur in any given circumstances. When one is talking *about* a theory—whether establishing it, or identifying a system as one to which it applies—one is concerned with what has been found to be the case, not with what must be; but when one is talking *in terms of* a theory—applying it to explain or foretell the phenomena occurring in such-and-such a situation—one is then concerned with what, according to that theory, must happen in that situation. There are several mistakes into which it is possible to be led if one fails to see just where it makes sense to say 'must', 'necessarily' and 'cannot', and where one has rather to say 'has been found'—one such is the kind of determinism which we shall have occasion to examine in Chapter V. These mistakes are, one finds, made only easier by the scientist's customary idioms ("If the wall is 6 ft. high and the sun is at 30°, the shadow *must be* 10½ ft. deep"), for in these the currently accepted theories of optics are employed without being explicitly mentioned. Logicians, for the sake of clarity, can afford to say the same thing less compactly but more explicitly, in onlooker's instead of participant's language: "If the wall is 6 ft. high, and the sun is at 30°, then a proper application of the theories of optics which *have been found* reliable in such circumstances as these will *necessarily* lead us to the conclusion that the shadow will be 10½ ft. deep."

What lies behind the Lockean view of laws of nature seems, then, to be their use as principles of inference: the necessity to which they point is the necessity with which conclusions follow when one argues in accordance with these principles. One may ask, then, why this necessity should seem 'opaque to the intellect', when principles such as that nothing can be both red and green all over are 'transparently necessary'. The subject is too large for us to go into it fully here, but perhaps a hint can be given. The difference seems to lie in this: we learn words like 'red' and 'green' at an early age, at the same time as we learn to sort, fetch, carry and label the things around us, and our knowledge that nothing can be both red and green all over is something which ordinarily shows itself in our ability to give and obey orders, and to make and understand reports, in which the words 'red' and 'green' appear. Only much later, when both the use of these words and the activities in connexion with which we have learnt to use them are second nature to us, do we come to ask why such a principle holds; and it then seems to us, naturally enough, that anyone who has got the hang of the words will recognize the force of the principle. In the case of laws of nature, on the other hand, one has neither the same strong association between the words appearing in the laws and those particular inferring techniques with which the laws belong, nor the same years of familiarity with the use of these techniques. As often as not, in fact, terms are taken over from outside physics and put to new jobs, and in consequence it may well seem far from obvious that 'light' must travel in straight lines, or that 'action' and 'reaction' must be equal and opposite. But perhaps, if dynamical calculation were second nature to us, in the way colour-classification is, and if we could all recognize, e.g., purely gravitational systems by eye, in the way we can tell red from green, the difference might not seem so great; and we might think the Law of Gravitation quite as transparent as the more familiar principles of colour-classification.

The point of Hume's 'constant conjunction' theory we have seen in part already: it is to rebut the suggestion made by advocates of the Lockean theory, that laws of nature somehow provide us with information about 'necessary facts'. (Recall also Mach's opposition to the idea that physics reveals necessities in nature.) In consequence, one finds Hume and his followers concentrating their attention, not on statements of type (iv), but rather on those in class (iii). "The light getting over this wall *is travelling* in a straight line," "The beams of light outside and inside this prism *are oriented* in such a way", "The salt *is dissolving* in this water"; these statements may constitute quite genuine applications of laws of nature, but there is nothing necessarily true about them. They just represent the sorts of thing that are in fact found to happen; and, by contrast with statements in class (iv), there is not even any 'must' in them. Of course, if one has a satisfactory theory to explain these facts, one will be able to show in any particular case that things must happen just as they are found to do: indeed, it would not be a satisfactory theory if one could not. But, to repeat, this is not to say that the facts explained are 'necessary facts': rather it is to say that they are necessary consequences of the theory. The distinction between necessary consequences and necessary propositions is obvious enough in elementary arithmetic: if a housewife argues, "I started with twelve pounds of sugar, and I've used four, so I must have eight pounds left," the formula on which she relies $(12 - 4 = 8)$ may be necessarily true—or rather, unconditionally applicable —but the conclusion she reaches ("I have eight pounds left") is to be accepted, not unconditionally, but rather as a necessary consequence of her data. The same thing holds in physics: when one applies a physical theory to a specific case, the conclusions to which one is led may, in the circumstances, be necessary ones, but it is a mistake to read this 'necessary' in the logic-book sense, as 'necessarily true'.

If there is no need for the Lockean and Humean views to be regarded as rivals, why then have they been so regarded? This will be clearer if we ask the question, "Are laws of nature necessary propositions or contingent ones?" For if we regard this dichotomy as exhaustive, and try to fit laws of nature into one category or the other, we shall find it hard to know what to say. Are we to say that, despite their empirical origin, laws of nature are necessary propositions? Or are we to say that, despite their claim to tell us what 'must' happen, they are only contingent propositions about constant conjunctions? Or must we contradict ourselves, by saying that they are both necessary and contingent? None of these alternatives is satisfactory, and the moral of our earlier discussion is that we should accept none of them. It is only because philosophers have come to laws of nature from such everyday statements as 'Rabbits are animals' and 'Rabbits eat lettuce' that they have supposed that laws of nature must be either necessary (like 'Rabbits are animals') or contingent (like 'Rabbits eat lettuce'). In fact, when they have attempted to establish their views that laws of nature are the one or the other, they have talked in either case, not about things which are properly called 'laws of nature', but rather about one or other of the types of statement which we have distinguished from the laws themselves.

Advocates of the 'necessary' view have, as we saw, paid special attention to those applications of laws of nature in which one is led to conclude, e.g., that a particular shadow must be 10 ft. 6 in. deep. But such a conclusion is not itself a law or principle, or a deduction from any law or principle: it is an inference drawn in accordance with the law or principle. The appearance in this statement of the word 'must', reflecting the use of a rule of inference, cannot therefore be taken as evidence that laws of nature are necessary propositions in any but a highly misleading sense.

Advocates of the 'contingent' view, on the other hand, have concentrated their attention, not on the laws of nature themselves, but upon the facts that they are used to explain—salt's dissolving in water, shadows being the depths they are, light-beams having the directions they do—all things which may with some justice be spoken of as regularities or constant conjunctions. But, once again, the statements they cite are not laws of nature at all, and again nothing is proved about the status of laws of nature by pointing to these facts.

In its way, to call laws of nature 'contingent' is as misleading as to call them 'necessary', for to do so is to focus too much light on a set of questions which never arise with reference to laws of nature, namely, questions about truth and falsity. It may be clear enough what it would mean to deny, e.g., that the law of gravitation applied to electromagnetic radiation, or again to deny that, the law being what it is, such-and-such a configuration of bodies must move in such-and-such a way; but it is quite unclear what it would mean to talk of denying the law of gravitation itself. One might say "It needs reconsidering and reformulating to fit it into relativity theory", but to say this is not to say that it is false: in such a case, the word 'false' cannot get a grip. The facts which scientists investigate experimentally have to do with the scope of their laws, and with what, applying the laws in a particular context, they must expect to happen. Physicists never have occasion to speak of the laws themselves either as corresponding or as failing to correspond to the facts. The logical relation between the laws and the facts is indirect: by talking as though they were connected any more closely than they are, one creates only confusion and misunderstanding.

3.8 *Whitehead and Schlick*: *Are laws of nature restricted generalizations or maxims?*

Where advocates of the first two views are preoccupied with statements of types (iii) and (iv), the 'restricted generalization' view seems to spring from a consideration of those in class (ii): i.e. statements about the discovered scope of laws of nature. As Kneale interprets him, Whitehead supposed that laws of nature must be generalizations of some kind, either restricted or unrestricted; and concluded, reasonably enough, that a few hundred years' experiments on this Earth could hardly justify us in advancing generalizations of a completely unrestricted kind. The natural consequence of this argument was the view that laws of nature are generalizations of a kind that tacitly refer to all places and times within a single, vast but bounded cosmic epoch.

Now there is an important point behind this account, but it needs re-stating. For, as stated, it assumes that the question to be asked is "Are laws of nature true always and everywhere?"; whereas the proper question is "Are laws of nature applicable equally at all times and places?" And the answer to the question is not "Yes, curiously and amazingly enough, they are found to be universally true," but "Yes, they are formulated in such a way as to be universally applicable: this is a feature which marks off laws of nature from the other statements of physical theory." If laws *were* universal empirical generalizations, it would indeed be a question whether they were always true; but they are not, and the point at issue must be put otherwise.

The heart of it can perhaps be illustrated in this way: one distinguishes in physics between those expressions which are to be labelled 'laws of nature,' and those expressions which are not so much laws of nature as applications of laws to special ranges of circumstances. Thus we can distinguish between the Law of Gravitation, a genuine law of nature, and such a statement as "Freely falling bodies accelerate by 32.2 feet/second every second": this latter expression is not itself a law of nature, but is an empirical law which can be accounted for by applying the Law of Gravitation to the special conditions of the Earth. Now it certainly makes sense to speak of our discovering that what we now call 'the Law of Gravitation' should itself be regarded as a law of this latter kind. This would happen if it were found, e.g., that over the whole region to which we had previously had access there was a constant 'field' of a hitherto-unrecognized type; and if, on investigating the properties of this field, we found that the law of gravitation could be expressed in its present form only for so long as this field remained constant. One can imagine, say, the value of the gravitational constant, 'G', being found to depend on the strength of this field. If this happened, we should have to reformulate our law so as to take account of the new discovery, and the present formula would be dethroned. The success of our present law would then be spoken of as a local and temporary consequence of the 'true law', in the way that the rate of gravitational acceleration on the Earth is now regarded as a local and temporary consequence of our present law.

This, however, does not prove that laws of nature tacitly apply only to limited regions of space and time, as it would if our Law of Gravitation were a simple generalization. On the contrary, the fact that such a discovery would be sufficient reason for dethroning our present law shows just the reverse: it shows that only those formulae we are ready to apply equally at all places and times qualify for the title of 'laws of nature'. But this, in its turn, does not imply that laws contain the words 'always and everywhere' in them either explicitly or tacitly. These words would be out of place within a law, and belong rather in statements of class (ii), about the circumstances in which any particular law has been found to hold. So Whitehead's suggestion, too, involves the confusion between laws and generalizations. Nor, for that matter, is the fact that it makes sense to say, "Perhaps our so-called Law of Gravitation is only a local affair", any reason for despondency: there is not the slightest reason at the moment to suppose the existence of the undiscovered field which would force us to this conclusion. Of course it *makes sense* to say, "Perhaps we have not got the law right". Nevertheless, we shall need to have good reasons before we abandon our present formulation of the law for another.

Finally, let us consider the view about laws of nature put forward by F. P. Ramsey, and quoted above in the words of Moritz Schlick: the view that such laws are not so much 'statements', 'assertions' or 'propositions' as 'instructions for the formation of propositions', 'rules of conduct', 'maxims' or 'directions for the investigator to find his way about in reality'. Again we shall find that the theory draws attention to something important about laws of nature, but once again this feature is described in a needlessly paradoxical way.

One can at any rate say in favour of this theory that its advocates are genuinely concerned with laws of nature (i.e. class i above), and not with those other, related classes of statement (ii, iii and iv) which have so often been confused with them. For the point which Schlick and Ramsey have wanted to emphasize is the one we ourselves have recognized as crucial: the fact that words like 'true', 'false' and 'probable' are applicable, not to laws themselves, so much as to the statements which constitute applications of laws; and that any abstract statement of a law or principle gives us only the form of a regularity, telling us by itself nothing about the phenomena it can be used to explain. As Schlick says, laws of nature "do not have the character of propositions which are true or false", and in some ways his alternative account of them is not at all a bad one. If we consider the techniques of geometrical optics, which give the Principle of Rectilinear Propagation its point, we can indeed see grounds for speaking of the principle as a means of finding one's way about in reality; and when we remember how far laws of nature are used as principles of inference, there is clearly some virtue in talking of them as rules for the formulation of statements about the world.

There is, in fact, only one thing about Schlick and Ramsey's account to which one can seriously object, and it is this same thing that gives the account its paradoxical air—the fact that they use unduly imperatival words such as 'instructions', 'directions' and 'rules', instead of some rather less exciting word such as 'principles'. If one makes this one amendment, the objections brought against their view, e.g. by Kneale, lose all their force. For Kneale argues that "if the sentence which purports to formulate a law gives [as Schlick suggests] only a general rule of conduct, what is derived from it can be no more than a command or injunction": as he sees it, on this view there would be no possibility of using a law to derive genuine propositions about the world—one could get only a string of particular injunctions. But Schlick and Ramsey are not claiming that laws of nature are generalized commands; the point of describing laws of nature in their way is to remind us of their use as inference-licences entitling us to argue from known facts about a situation to the phenomena we may expect in that situation; and the weakness of Kneale's objection becomes clear if one considers how his argument would affect other principles of inference.

Consider, e.g., the Principle of the Syllogism. Lewis Carroll showed in his paper, *What the Tortoise said to Achilles*, what impossible conclusions one is led into if one treats the Principle of the Syllogism as a super-major premise, instead of as an inference-licence; yet it does not follow from his discovery that the conclusions of all valid syllogisms, which may loosely be spoken of as 'derived from' that Principle, must therefore be commands or injunctions. This would be the case only if one confused conclusions deduced from the Principle with those inferred in accordance with the Principle: the phrase 'derived from the principle' hides this distinction. It is the same with laws of nature. The conclusions about the world which scientists derive from laws of nature are not deduced from these laws, but rather drawn in accordance with them or inferred as applications of them, as our examples have illustrated. It is only if one takes Schlick's phrase 'rules of behaviour' too seriously that Kneale's objection carries weight. Regarded as principles of inference—though ones whose range of application is empirically bounded—laws of nature do indeed have very much the sort of job that Schlick attributes to them. Certainly they act hardly more as premises in physical arguments than the Principle of the Syllogism does in syllogistic ones.

What is it that makes Schlick's way of putting his thesis especially paradoxical? It is perhaps this: that it snaps the link between laws of nature and the world. Like the phrase 'laws of our method of representation', Schlick's phrase 'directions for the investigator' seems to sever laws of nature from the world entirely, and makes it appear that they have to do solely with physicists and their conduct. But to snap this link is, as we saw earlier, an extremely misleading thing to do. "Through their whole logical apparatus the laws of physics still speak about the objects of the world"; and the fact that some inferences rather than others come to be licensed usually tells us much more about the world than about the physicist and his methods. (Though this is not equally so in every case, as will be seen when we discuss Eddington's views on the subject in the next chapter.)

How are we to account for Schlick's choice of this unhappy form of words? The reason for it seems, strangely enough, to be the same as that which distorts the Lockean and Humean views—the assumption that the only statements representing genuine 'propositions' are those which are straightforwardly classifiable either as necessary or as contingent. Where the 'principles of necessitation' view classes laws of nature as opaquely necessary propositions, and the 'constant conjunction' view classes them as contingent propositions of a somewhat sophisticated kind, Schlick sees the unsuitability of putting them in either category. But his reaction is too strong. For his conclusion is that, if laws of nature are neither necessary propositions nor contingent ones, they cannot properly be spoken of as propositions at all: they must accordingly be found a place with those other alleged quasi-propositions, the pre-scriptions and recommendations of ethics and aesthetics. Hence the imperatival words he chooses: 'instructions,' 'directions' and 'rules of behaviour'. As so often in philosophy, in objecting very properly to his opponents' conclusions, he is betrayed into the same fallacy as they.

Schlick talks of the investigator finding his way about in reality, Ryle of law-like statements as inference-tickets. Perhaps these metaphors can be combined. For there is one variety of railway ticket not unlike laws of nature—the 'runabout ticket'. Tickets of this kind do not have a single starting-point and destination printed on them: they are valid, instead, for an unlimited number of journeys within a given stretch of country. The extent and limits of this region need not be, and usually will not be stated on the ticket: they will be specified elsewhere —e.g. on posters—and they can be varied by the railway authorities without the ticket looking any different. Now one might buy one of these tickets without knowing what its region of validity was; but one could find this out experimentally, by seeing at what stations it was accepted. And one can do worse than think of the physicist as a man who, in formulating laws of nature, prints his own runabout tickets, and thereafter makes it the goal of his experiments to discover where he can get with their help. The formal statement of a law is like the runabout ticket itself, which shows on it nothing as to its scope: it is as a result of experience that the physicist comes to know within what region it can be confidently employed.

By making the journeys (inferences) so licensed, the physicist finds his way around phenomena: by thinking of the systems he studies in terms of appropriate models, he *sees* his way around them and comes to understand them. But there is one important preliminary—first he must be able to *identify* each system, classify it in theoretical terms, recognize its place on the map. As we shall have reason to emphasize in Chapter V, this is a logically vital step; and it is by no means as trivial as one might think. Physical systems do not carry identification labels, as railway stations do; nor is there any way in which they can tell us themselves where on the theoretical map they belong. Anyone who has studied chemistry will know what a business identifying an anonymous specimen can be. What still needs to be recognized is the logical burden which the task of identification is made to bear.

CHAPTER IV

THEORIES AND MAPS

WE have seen how natural it is to speak of ourselves 'finding our way around' a range of phenomena with the help of a law of nature, or 'recognizing where on the map' a particular object of study belongs. In doing so, we are employing a cartographical analogy which is worth following up; for whereas to treat laws of nature on the pattern of generalizations is positively misleading, and to think of them as rules or licences reflects only a part of their nature, the analogy between physical theories and maps extends for quite a long way and can be used to illuminate some dark and dusty corners in the philosophy of science. Of course, like any analogy, it will take us only a certain way, but after an overdose of arguments in which physics is treated on the pattern of natural history, it can act as a healthy purge. That this should be so is no accident, since the problems of method facing the physicist and the cartographer are logically similar in important respects, and so are the techniques of representation they employ to deal with them.

4.1 *Ray-diagrams and equations as maps of phenomena*

Let us return, as a first application of this analogy, to a question we considered in an earlier section. This is the question the phenomenalists tried to answer: how we are to think of the relation between a scientist's experimental observations, all of which are expressible in everyday language, and the corresponding theoretical statements in which the technical terms of the science appear.

The difficulty to be overcome before we can answer this question arises as follows. Mach wanted to insist, rightly, that a scientific theory draws its life from the phenomena it can be used to explain: furthermore, the idea that the scientist needed insight into the causal connexion of things smacked to him of metaphysics, and he tried to do without t. In view of this, it was natural for him to suppose that, if a law of nature was to contain no more than the phenomena it was used to explain, it must be thought of as a summary of them, i.e. as an abridged description or comprehensive and condensed report of the experimental observations: "this," he concluded, "is really all that laws of nature are." But such an account of the matter may get us into difficulties. For to speak of laws as condensed summaries, abridged descriptions, or comprehensive reports, suggests that the connexion between any set of experimental observations and the law they are used to establish is a *deductive* one, so that it should be possible to give mechanical directions for producing a theory from a set of observations, much as one can produce a statement about the average schoolgirl from a set of measurements of individual schoolgirls. And this, as we saw, is a mistake: the relation between laws and phenomena cannot be so described.

How then are we to restate this connexion without abandoning the ground Mach gained? This is where the analogy between theories and maps can help us, for a simple cartographical example will show that no deductive connexion need be looked for.

Consider, for instance, the imaginary motoring map opposite, showing the town of Begborough and its environs.

We can ask about this section of map a question similar to Mach's question: namely, what relation it bears to the set of geographical statements that can be read off it, such as "Potter's Bridge is 5 m. NE of Begborough on the road to Little Fiddling", and "Great Fiddling is 3 m. due West of Little Fiddling."

How are we to answer this question? Certainly the map cannot be said to be deduced from the set of geographical statements nor, in a logic-book sense of the phrase as opposed to a Sherlock-Holmesian one, are the statements deduced from the map. For in a deductive inference, such as "Fish are vertebrates, mullet are fish, so mullet are vertebrates", the same terms appear both in the premises and in the conclusion; whereas here the 'conclusions' read off may be statements, but the 'premise' is a map and contains no 'terms' at all. Only where premises and conclusion are comparable in the way that "Fish are vertebrates" and "Mullet are vertebrates" are comparable, is there room for a deductive connexion, so the relation between the map and the geographical statements must be of a different, non-deductive kind. At the same time, the map need not be said, in Mach's sense, to 'contain' anything which cannot be expressed as a geographical statement of the kind included in our set: everything which one could read off from the map is of this sort. Though the map and the geographical statements are not deductively related, one need not

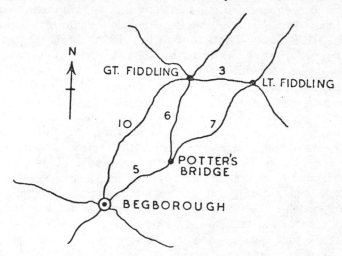

conclude that the map goes beyond the surveyor's readings; since it does not present us with additional information of a novel kind, but represents the same information as the statements in a different manner. This example shows that, when we are presented with two logically incomparable forms of expression, the question whether or no one form of expression contains more than the other is quite independent of the question whether or no the one can be deduced from the other. In fact, unless the expressions are of logically similar kinds, there can be no question of such deduction.

The logical relation between, for instance, ray-diagrams in geometrical optics and the phenomena they can be used to represent, is a similar one. Here, too, neither can be spoken of as being deduced from the other: yet a ray-diagram need not be thought of as containing more than the phenomena. It is rather that the diagrams present all that is contained in the set of observational statements, but do so in a logically novel manner: the aggregate of discrete observations is transformed into a simple and connected picture, much as the collection of readings in a surveyor's note-book is transformed into a clear and orderly map.

The consequences of this analogy are worth noticing. For if someone asks, "Doesn't the map tell us that Potter's Bridge is 5 m. NE of Begborough, and a whole lot of similar things?", we can only answer "Yes and No." Certainly, if you know how, you can *read off* from the map a great range of geographical information; but the map on the one hand, and the geographical statements on the other, tell us things in very different ways. A man might own Ordnance Survey maps of the whole country, and yet, for lack of a training in map-reading, be quite unable to tell us anything of a geographical kind: likewise, a man might have memorized all the currently accepted laws of nature and even know a vast amount about the calculative side of mathematical physics, and yet not be equipped to explain or predict any of the phenomena observed in the laboratory. The most the first man could do would be to lend the appropriate map, on request, to a man capable of reading it: in physics, too, the mathematician remains the servant of the man who knows when and how the results of his computations can be applied. Jeans and Eddington were both primarily mathematicians, and in their popularizations of physics gave prominence to the mathematical side of the subject, but the results were in certain respects misleading: the physics is not in the formulae, as they suggested and as we are often inclined to suppose, any more than being able to find your way about is part of a map. The problem of *applying* the theoretical calculus remains in physics the central problem, for a science is nothing if its laws are never used to explain or predict anything.

To pursue our analogy yet further, we may ask: if the map and the ray-diagram are counterparts, and the observations of the surveyor and of the experimenter are also counterparts, what exactly corresponds in cartography to laws of nature in physics? Here the analogy begins to fail us, for interesting reasons. For to press it at this point would mean saying that laws of nature in physics were to be thought of as the counterparts of the laws of projection in accordance with which one produces any specific type of map, such as Mercator's; and this leads to difficulties.

In certain respects the parallel holds: we have already seen the parts the Rectilinear Propagation Principle and Snell's Law play in the production of ray-diagrams, and the laws of motion in dynamics play a similar part when one constructs the equations of motion of a dynamical system. Up to a point, therefore, the analogy with the laws of projection can be illuminating. But the comparison is also an unhappy one. The problems facing a cartographer have certain important common features. In each case, it is his task to represent a part of the surface of the Earth on a plane sheet of paper, so as to preserve certain chosen features, such as equality of area; and, the shape of the Earth being what it is, the rules of projection are calculable from his knowledge of the conditions of his task. But in physics the situation is very different. Though in some cases we may eventually come to be able to work out what form laws of nature will take, as when one derives the laws of geometrical optics from a knowledge of physical optics, this knowledge is not like the prior knowledge of the problem which we have in cartography.

In general, there seems to be no way of saying beforehand what sort of techniques of explanation will be appropriate in a given field of study. That is why laws of nature have always to be *discovered* in a way in which the laws of projection do not need to be. Our analogy could be preserved only by imagining the figure of the earth to be both irregular and discoverable only in the course of our cartographical survey if it were so, cartographers would be unable to pick on a method of projection beforehand, and would have to find out empirically, as they moved from region to region, in what manner each new area was to be mapped. Establishing a law by appeal to the results of experiment would be like showing that a satisfactory map of the new area could be produced using such-and-such a method of projection—as indeed we saw in the case of Snell's Law. But even when so amended, the analogy has its limitations: the problems to be tackled in physics differ widely from one another in a way in which problems of mapping can never do.

4.2 *The physicist as a surveyor of phenomena*

In the traditional logical account of the sciences, one encounters certain difficulties when explaining how it is that experiments are used to establish theories. In the first place, physicists seem to be satisfied with far fewer observations than logicians would expect them to make: one finds in practice none of that relentless accumulation of confirming instances which one would expect from reading books on logic. This divergence is partly to be accounted for by the logicians' confusion between laws and generalizations—one would hesitate to assert, say, that all ravens were black if one had seen only half a dozen of the species, whereas to establish the form of a regularity in physics only a few careful observations are needed—but this is not the whole story. There is also a second, related difficulty to be overcome: that of explaining how subsequent applications of a theory are related to the observations by which the theory was originally established.

To take the two difficulties together: it is worth noticing that they arise for theories as much as, and no more than, for maps. Not all the applications to which a theory is put need have been specifically made in the course of the experimental investigation by which it was established. But nor need all the things which can be read off from a map have been specifically put in. A child might wonder how it was possible ever to produce a map at all, since to tread every inch even of a small area, and to measure all the distances and directions that one can read off from a map, would take an unlimited length of time. This, of course, is the marvel of cartography: the fact that, from a limited number of highly precise and well-chosen measurements and observations, one can produce a map from which can be read off an unlimited number of geographical facts of almost as great a precision. But it is not a marvel calling for a general explanation, for only in some regions can the techniques be implicitly relied on. In irregular country it is always possible to be misled, and the number of observations which have to be made per square mile will be much greater in some areas than others—just how many are needed being something the practising cartographer must be able to judge.

Correspondingly, it is a fact that many physical systems have been found whose behaviour can be similarly 'mapped'. Having made a limited number of highly accurate observations on these systems, one is in a position to formulate a theory with the help of which one can draw, in appropriate circumstances, an unlimited number of inferences of comparable accuracy. Thus it is always possible that the next time Boyle's Law is applied, the particular combination of pressure and volume concerned will be being observed for the first time. But again, though this fact is in its way a marvel, it is not one requiring a general explanation, any more than is the possibility of mapping. For here, too, how far the behaviour of a given system consists of phenomena which can be mapped in a simple way, and just how many observations will need to be made before we can be confident that our theory is a trustworthy one, are things which will vary very much from system to system and which it is part of a physicist's training to learn to judge.

The difficulties which logicians find in understanding the role of experiments in physics arise, therefore, not only from their thinking so much in terms of generalizations: to get clear about it calls for quite a detailed study of the logic of physics. To put our point succinctly—only when a regularity has already been recognized or suspected can the planning of an experiment begin: until that time, the mere multiplication of experiments is comparatively fruitless. And when that time comes, the problem for the physicist will not be like that for the botanist or the naturalist, as it would be if his sole aim were to generalize about 'all lumps of rock' or 'all flames'—that is, if physics were the natural history of the inanimate. His problem will rather be like the surveyor's problem, and the accumulation of observations in large numbers will be as much a waste of energy in physics as in cartography. Faced with the demand for more and more observations the surveyor and the physicist might equally reply, "What's the point? We've been over that ground already."

There is a further point about the sorts of observations which need to be made in order to put a physical theory on a satisfactory footing. Logicians have remarked, rightly, that physicists prefer to make a limited number of observations covering a wide range of circumstances, rather than a larger number of observations covering a smaller range of circumstances. The point of this preference, they have concluded, must be to show that the laws being established are true generally, and not only true on certain conditions. From this point they have gone on, first, to develop an elaborate theory of confirmation, analysing the way in which conditional clauses might be eliminated from a hypothesis by reference to experimental data; and secondly, to formalize the process of theory-establishing in a way intended to fit in with the mathematical theory of probability, the aim being to find a way of assessing in numerical terms the probability of a given physical theory.

This account does not fit in with practice, nor does it properly explain the preference for varied observations. For physical theories are not spoken of in practice as true, false or probable, nor is it clear what one could be expected to understand by the statements, "The probability of the kinetic theory of gases is $\frac{17}{18}$" and "Five to one on Snell's Law". The point of varying the conditions of observation is, in fact, otherwise: it is to discover the *scope* of the theory, not its degree of truth or the conditions on which it can be accepted as true. The 'logic of confirmation' and the application of the probability-calculus to theories have, therefore, hardly the slightest relevance to the physical sciences. The mathematical theory of probability has some place in the process of theory-establishing, certainly; but it is a more restricted one than logicians have thought.

It has a central place only in limited branches of theory, such as statistical mechanics and parts of quantum mechanics: more generally, it has to do solely with questions of the form, "Can such-and-such a specific set of experimental observations be satisfactorily accounted for by applying a given theory in a given manner?"—i.e. the question whether the scatter in our observations is significantly greater than the probable errors in our measurements would lead us to expect. The application of the calculus of probability in this sort of way raises no general questions of a philosophical kind, but only particular questions of statistical technique: questions to be answered in terms of the theory of curve-fitting, significant deviations and so on.

4.3 *Degrees of refinement in cartography and physics*

There are many places in the physical sciences where one finds a single field of phenomena covered by two or more theories, in which techniques of different degrees of sophistication are employed. The optical phenomena with which we have been concerned are a good example. We saw earlier how the range of application of geometrical methods of representation in optics is restricted by diffraction and the like, so that the limited success of geometrical optics becomes itself something requiring explanation. To explain the phenomena that cannot be accounted for within geometrical optics, the wave-theory of light was introduced, and this theory was particularly acceptable because it could also be used to account for all the phenomena that geometrical optics covers. It is true that what is simple in the more elementary theory, explaining shadow-casting, for instance, tends to become laborious in the more refined one; but since the wave theory can not only be used to explain a wider range of phenomena, but does so to a higher degree of accuracy, and also explains just why the methods of geometrical optics break down where they do, it is accepted as providing a more fundamental explanation than the simpler geometrical account— and reasonably enough.

Where there is such a multiplicity of theories, certain things may appear mysterious to the outsider or to the beginner. What is the relation of the two theories to one another, and how does the development of the more refined theory affect the status of the simpler one? Does the change-over mean that the earlier theory has in some sense been falsified? If that is so, surely it should be regarded as discredited; so how is it that lens-designers, for instance, may prefer to go on using the geometrical techniques of ray-tracing after the wave-theory has been shown to be the true theory? Perhaps the most puzzling thing is the way in which notions which were central in the simpler theory—that of a light-ray, for instance—may disappear almost completely in the more refined theory. So long as we think in terms of the geometrical account, the term 'light-ray' is indispensable: light-rays indeed seem to be the principal actors on the optical stage. Yet in the wave-theory a light-ray is an artificial construct as compared with, say, a 'wave-front'; and Snell's Law, which is stated, as we saw, in terms of *rays* of light, has to be reformulated in quite a different way before a niche can be found for it in the new theory. Yet the phenomena are as they always were: lamps burn as they did, shadows fall where they did, rainbows, reflections and all are as they were. What then has happened to the light-rays?

The best answer can perhaps be given by pointing out first the relation between different types of map. The imaginary road map of the region between Begborough and the Fiddlings which we discussed a few pages back, need not be the only map of the region. There will also be some more elaborate physical maps drawn to a larger scale and showing a great deal more detail. In such maps as these, roads will perhaps be drawn to scale, not represented by lines of purely conventional widths, while towns and villages will be marked, not as mere dots and blobs of standard sizes, but as having definite shapes and made up of individual streets and blocks of houses.

Now a number of things should be noticed about the relation between the road map and a physical map of the same region. In the first place, many things can be mapped on the physical map which there is no way of putting into the road map: this is a consequence of the ways in which the two maps are produced, and of the comparative poverty of the system of signs used on the road map. On the other hand, given the physical map, one could produce a satisfactory road map: all that appears on the road map has its counterpart on the more elaborate map, even though in a different form. But this does not mean that the road map is not, of its kind, an unexceptionable map of the region. Providing that it is not thought of as having irrelevant pretensions, there is nothing wrong with it: indeed, for some applications one will be able to discover the things one wants to know, e.g. distances by car, more easily from the road map than from the physical one. Finally, it is worth noticing what happens if we mix up the systems of signs used on two different kinds of map. There are some motoring maps in which one finds town-outlines and other features sketched in on top of the simple road pattern: but since only distances along roads can be given a satisfactory interpretation on such maps, the result is usually confusing, and the simple blob for a town is more consistent with the general scheme of the map.

The relation between geometrical optics and the wave-theory is not unlike that between a road map and a detailed physical map. Thus the fact that one can explain on the wave-theory, not only all the phenomena that can be accounted for on the geometrical theory, but also why the geometrical account holds and fails to hold where it does, is like the fact that one can construct a road' map from a physical map; but again it is not a sign that the geometrical theory need be superseded for all purposes. Road maps did not go out of use when detailed physical maps were produced. It shows only that, as one can produce a road map from a physical one but not *vice versa*, so one could produce a ray-diagram from the wave-theory picture of an optical system, but not *vice versa*. The conceptual equipment of the geometrical theory, like the system of signs on a road map, is too poor for one to do with it all that can be done with the wave-theory. Indeed, the notion of a light-ray is an artificial one in very much the way that the conventional-width road is, and has to be abandoned in the wave-theory because the accuracy with which one wants to answer questions about optical phenomena is too great for the conventional picture to be retained. No more can one, from a simple motoring map, answer questions about the distance from the northern verge of one road to the middle of another —these are things that a map of that type does not pretend to show. Again, since there is no room within geometrical optics for representing the phenomena of diffraction, a physicist would hardly think it worth while to give any indication on a ray-diagram of the shapes of any diffraction-fringes he observed: they would be just as out of place there as town shapes are on a bare motoring map.

If we look at the relation between different theories from this angle, we can notice some points of importance about the notion of a 'fundamental' or 'basic' theory. One finds that, at a given stage in the history of physics, there is commonly one theory, at any rate in a particular field, which is regarded as the basic theory: this theory is thought of as capable of accommodating all the phenomena to be observed in that field. Now two questions need to be asked. Since it will never be the case that all the phenomena have in fact been explained, all that need be claimed is that the basic theory can in principle explain them all: the first question is, what are we to understand by this claim? Secondly, when physicists talk about explaining everything, what are the criteria by which they would judge that everything had in fact been explained?

It is helpful to compare the basic theory with the fundamental map on which the Ordance Survey might record all the things which it is their ambition to record. This would, of course, be a map drawn on the very largest scale, but it would not be the only true map of the country: rather it would be the one which most fully and precisely represented the region mapped, and the one from which by appropriate selection and simplification all others could be produced. For many purposes it will be too elaborate to be of practical use, but for some purposes none else will do, and the lover of cartography for its own sake must have a special place for it in his heart.

The value of the comparison lies in this: it suggests that the standards of what constitutes a complete theory in physics may change. For we could say that the fundamental map was complete only if it showed all the things which in that region it was the cartographer's ambition to record. Now it is always possible for cartographers to develop fresh ambitions: the criteria of the completeness of a map are, accordingly, at the mercy of history. So are they with the theories of physics. One is at first inclined to suppose that the physical sciences have a definite goal, the same for Aristotle, Newton, Laplace, Maxwell and Einstein, but a closer look at the history of the subject will show the mistakenness of this idea. Rather there is at any given stage a standard of what sorts of things require explaining: this is something with which scientists grow familiar in the course of their training, but which is hardly ever stated. The standard accepted at any time determines the horizon of physicists' ambitions at that time, the goal which for them would have been reached if 'everything'— i.e. everything thought of as requiring explanation—had been found a place in the theories of physics.

In physics, as in travelling, the horizon shifts as we go along. With the development of new theories new problems are thrown into prominence, ways are seen of fitting into physical theory things which before had hardly been regarded as matters requiring a place at all: the horizon accordingly expands. Classical physics, for instance, was thought of as potentially exhaustive. Yet, looking back at it, we must feel that nineteenth-century standards of exhaustiveness were strangely unexacting. The existence of ninety-two elementary kinds of matter, their relative abundance, and the colour of the light emitted by each element: these things, to mention only a few, were hardly even asked about. They were not things to be explained but things to which, in a phrase of Dr. Waismann's, 'one had to take off one's hat'. Perhaps this is why the claim of some classical physicists, that they had the explanations of everything in principle in their grasp, was peculiarly distasteful. For what was repugnant was not just the fact that the theories advanced were so bare and mechanical but, quite as much, the fact that their idea of what it would be to have explained everything was so much smaller than life.

On the whole, then, the horizon of physics expands. From time to time, however, the ideal changes in a way which cannot be described so simply, and these are occasions when disputes of a philosophical kind frequently arise. In the change-over from Aristotelian to Newtonian dynamics, for example, certain phenomena which were previously regarded as 'natural' and taken for granted, such as carts stopping when the horses ceased to pull and heavy bodies falling to the ground, came to be thought of as complex phenomena needing explanation: in these respects the horizon expanded. But at the same time certain other phenomena, which had until then been regarded as complex and in need of explanation, were reclassified as simple, natural and to be taken for granted; notably, the continued flight of an arrow after it had left the bow, and the unfaltering motion of the planets along their tracks. The need for this second kind of reclassification was the great obstacle to the development of the new dynamics: it was easy enough to recognize as complex something previously accepted as simple, but the reverse change was a bitterly hard one to make. And so it has been elsewhere. One finds the same thing happening around 1700, in the dispute between Leibniz and the Newtonians over the mechanism of gravity and action at a distance; and the same thing again in the late nineteenth-century disputes over the luminiferous ether.

One of the most instructive disputes of this kind is in progress at the moment, and concerns the adequacy of quantum mechanics as a basic theory. Einstein, on the one hand, refuses to accept the changes in our standards of what needs explaining which have to be made when one introduces quantum mechanics: in his view, these changes require one to restrict the horizon of scientific endeavour in an unjustifiable way. His opponents, on the other hand, claim that his objections show only that he has not properly understood the theory. This is not the place to deal with the substance of the dispute. But for our purposes the language in which the dispute is carried on must be noticed; for it is couched in terms of the question, "Is a quantum-mechanical description of a physical system complete or not?" This way of posing the problem confuses the issue, giving to it too sharp an appearance of opposition. For a complete description of a physical system is one from which one can, using the currently accepted laws of nature, infer all the properties of the system for which it is a physicist's ambition to account: where two physicists do not share a common standard of what does and does not need to be explained, there is no hope of their agreeing that the corresponding description can be called complete. This is the case in *Einstein v. the Rest*: the use of the word 'complete', with its implicit reference to particular criteria of completeness, conceals rather than reveals the point at issue between the parties. A similar moral holds more generally: by using the words 'exhaustive', 'all', 'everything' and 'complete' in stating the goal of their investigations, physicists have hidden from themselves as well as from others the changes in the horizon towards which they work.

4.4 *Causes are the concern of the applied sciences*

A subject which receives a good deal of attention in traditional treatments of 'induction and scientific method' is that of causes. It is a proposition often taken as obvious that the task of the sciences is the discovery of causes: Mill's four methods and similar formal analyses can, indeed, be regarded as relevant to the physical sciences only in so far as this is so. Some logicians go further: the existence of causal chains is said by them to be a condition of the possibility of science, and certain features of quantum theory are accordingly interpreted as a breakdown of the causal principle or an abandonment of causality. Causes, causation, causality: these are the staple of much philosophical and logical writing about the sciences.

If one turns from the logic-books and the spare-time philosophical works of scientists, to the professional journals in which the sciences really progress, one is in for a surprise. For in the papers there printed the word 'cause' and its derivatives hardly ever appear. In works on engineering, perhaps; in medical journals, certainly; wherever the sciences are applied to practical purposes, there one finds talk of causes and effects. But in the physical sciences themselves, the word 'cause' is as notable an absentee as the word 'true'. Why is this?

To recognize the reason, consider first the sorts of everyday situation in which we have occasion to ask questions about causes. A wireless set, instead of giving out a Haydn symphony, howls dismally; an invalid's temperature, instead of staying at 98.4°F., soars to 105°F.; a stretch of railway embankment crumbles and falls away, leaving the lines in a dangerous state; a field of barley grows unevenly, sturdy and thick in some parts, sparse and weak in others; and in each case we ask about the cause—why the wireless does not work properly, what is wrong with the invalid, what has happened to make the embankment collapse, in what respect the fertile parts of the field differ from the infertile ones. Developments which we are interested in producing, preventing or counter-acting—these are the typical sorts of thing about whose causes we ask. Correspondingly, to discover the cause of one of these developments is to find out what it is that needs to be altered, if we are to produce, prevent or counteract it. To discover the cause of the howling is to discover, say, that a particular valve is faulty and needs replacing; the patient, it may be, has an infected sinus; the foundations of the embank-ment have been sapped by an underground stream; the fer-tility of the different parts of the field depends upon their nitrogen content. In each case, we speak of that as the cause which *in the context* would have to be, or have had to be other-wise, for the development on which our attention is focused to go differently. Where there is no one thing in the antecedents rather than anything else which could reasonably have been wished different, we may accordingly find no use for the term: "Nothing particular caused it", we sometimes say, "Things just worked out that way".

Now these everyday cases are all anthropocentric: the things for whose causes we seek are those we human beings want to produce, prevent or counteract. Our examples of typical everyday uses of the term 'cause' are, that is, all concerned with *people getting somewhere*. That they should be anthropocentric is not essential. One can ask about the cause of the explosion of a distant star as well as of an invalid's temperature: things which, humanly speaking, are indifferent can have causes just as well as things we care about. But one feature of our examples is essential. Wherever questions are asked about causes, some event, which may matter to us or may not, has a spotlight turned on it: the investigation of its causes is a scrutiny of its antecedents in order to discover what would have to be different for this sort of thing to happen otherwise—what in the antecedents God or man would need to manipulate in order to alter the spot-lighted event. It is not essential that the search for causes should be anthropocentric; but that it should be *diagnostic*, i.e. focused on the antecedents in some specific situation of some particular event, is essential. People sometimes mystify one by asking what would happen if the order of all physical events were reversed, suggesting that as a result effects would then precede causes. This suggestion misses the point of the notion of cause—in particular, its dependence on context. If one puts a steam-engine into reverse, one has to apply the brake at quite a different point in the cycle in order to achieve a given result, e.g. to stop it at top dead centre: in the new context the same pairs of happenings no longer belong together as causes and effects. But the causes are still, necessarily, among the *antecedents* of the effects.

It is, then, still in cases where our interest is in how one might 'get somewhere', i.e. produce or counteract some spotlighted development, that we talk about causes—though the destination need not be one that we care about either way. From this we can see why the term 'cause' is at home in the diagnostic and applied sciences, such as medicine and engineering, rather than in the physical sciences. For the theories of the physical sciences differ from those of the diagnostic and applied sciences much as maps differ from itineraries. If the term 'cause' is absent from the physical sciences, so also a map of South Lancashire does not specifically tell us how to get to Liverpool. To a man making a map, all routes are as good as each other. The users of the map will not all be going the same way, so a satisfactory map is route-neutral: it represents the region mapped in a way which is indifferent as between starting-points, destinations and the like. An itinerary, however, is specifically concerned with particular routes, starting-points and destinations, and the form it takes is correspondingly unlike that of a map. Often enough, of course, a map may be used to work out the itinerary for a particular journey, and from one map an indefinite number of routes may be read off, as occasion requires. But, from its form, there is nothing about a map to show that it is to be used for this, rather than any other of a wide range of purposes.

In the physical sciences, likewise, the regularities we find in any particular field of phenomena are represented in a way which is application-neutral. The theories which are produced to explain optical phenomena, for instance, do not specifically tell us how to bring about this or that optical effect—how to produce a shadow a hundred feet deep, or how to create a mirage. Rather they provide us with a picture of the sorts of phenomena to be expected in any given circumstances, which can then be used in any of a number of ways. The study of the causes of this or that event is, therefore, always an application of physics. It is not of direct importance to the physicist, and can at best suggest to him something which may turn out to be of theoretical importance. In the case of theories, as of maps, there will be an indefinite number of applications to be made, say, in engineering. But the way in which the theory is formulated will not show that it is to be applied in this or that particular kind of way, for the production or prevention of this or that particular kind of development. Problems of application and questions about causes arise with reference to particular contexts, but physical theories are formulated in a manner indifferent to particular contexts: it is when we come to apply theories that we read off from them the causes of this and that, but there is no call for the term 'cause' to figure within the theories themselves.

This analogy shows us something about the relation between the fundamental and applied sciences, and about such phrases as 'applied physics'. For in many fields of science practical skills preceded theoretical understanding, and even provided the first data for systematic study. Sundials were in use for centuries before their operation was properly understood, and there are still plenty of familiar processes, in cooking for instance, about whose physico-chemical nature we have only the sketchiest of ideas. There is therefore only a part of engineering which can be called 'applied physics', even though this part may be continually growing and may in some divisions, such as atomic energy, be all but exhaustive. This state of affairs also has its natural counterpart in cartography. For a long time, travellers relied on itineraries rather than on maps; Greek seamen and Roman legionaries as often as not followed set routes for which itineraries had been written out; there must still be today a few more remote parts of the world which are totally unmapped, but around which a guide could take one; and even in our own well-mapped country we all know some short cuts and refinements that are shown on no map. So though the preparation of itineraries may in fact often be applied cartography, it need not be. Itineraries preceded maps. The development of cartography has given us a way of understanding the relations between different routes, and at the same time a source of new itineraries whose possibility had not previously been recognized. And there may be some parts of the world so remote, so mountainous, that one could hardly hope to work out itineraries for them except by first mapping them from the air.

The absence of the term 'cause' from the professional writings of physicists can therefore be explained. But this explanation in its turn creates a fresh problem: for if the prime aim of the physical sciences is not the discovery of causes or causal chains, what are we to make of the elaborate discussions of causality and indeterminacy provoked, e.g., by quantum mechanics? The subject is too complex to go into in detail here. But one thing may be worth saying: the idea of causality reigning unchallenged seems to be accepted by philosophical scientists so long as the basic theories of the time appear capable, in principle, of explaining all the things it is hoped eventually to explain. It is no surprise, accordingly, to find Einstein, whose horizon stretches further than quantum mechanics can reach, calling for a re-establishment of causality, and saying reproachfully that Born and his colleagues 'believe in a dice-playing God'. Restated in our terms, the question of causality becomes the question whether all physical phenomena are completely mappable; and this, like other general philosophical questions containing the words 'everything', 'all' and 'complete', depends very much on one's standards of completeness. The determinate, correspondingly, is that for which a place can be found on the map; so that the very name 'Indeterminacy Principle' for Heisenberg's relation seems to rest on a misunderstanding.

The notion of causal chains and causal contiguity, which Russell for one regards as central for the justification of scientific method, must wait for a proper discussion till we consider determinism and the 'causal nexus' in Chapter V, but again a word is in place here. The idea that events form chains, each drawing the other inevitably after it, originates in what we have called the diagnostic field rather than in the physical sciences. It is catastrophes of which we most want to know the causes, and the discovering of such a cause is spoken of as 'laying bare the chain of circumstances which led to the disaster'. Two things must now be noticed. First, the idea of a chain of circumstances tends to be taken too seriously on such occasions just because it is a disaster whose causes we are concerned to diagnose, i.e. the sort of thing we tend also to think of, as often as not mistakenly, as fated or destined to happen: apart from this association, there is no reason to understand the 'chain' metaphor as any more than a metaphor. Secondly, this tendency is reinforced by special features of the diagnostic, as opposed to the physical sciences. To understand the causes of something is the first step towards being able to cause it to happen. Success in the applied sciences may therefore lead us to think of events as at the ends of chains; all we need is to know which chain to pull and the required result will follow. But simple chain-like prescriptions can be given only in restricted sets of circumstances: we can confidently match causes and effects only in a given context. So once we shift from the diagnostic to the physical sciences the idea of a causal chain is of as little use as the term 'cause' itself.

4.5 *Eddington and the fish-net*

A perplexing question about the theories of physics was raised by Sir Arthur Eddington, and has been widely discussed in the last few years. "How much," he asked, "of the structure of our theories really tells us about things in Nature, and how much do we contribute ourselves?" This question was of importance to him because of his own professional activities, for it was his aim to 'work out from first principles', and treat as a conceptual matter, quantities which many of his fellow-physicists regarded as matters of brute fact. One instance is the ratio of the mass of the proton to that of the electron, a quantity which many physicists regard as something to be discovered only by looking and seeing, like the ratio of the populations of London and Liverpool: another is the number of protons and electrons in the Universe, which Eddington regarded as a conceptual matter but his critics as a pure matter of fact, like the aggregate population of the Earth.

Now there is an important philosophical question here, which is worth a more careful examination than it has so far received. Much of the discussion that it has been given has been needlessly mystifying, and some of it is completely misconceived. The conclusion has even been drawn from Eddington's suggestions that the theories of physics are essentially subjective—imposed on the facts, even to the extent of falsifying them, rather than built up so as to give a true picture of them. One is reminded of Bergson's thesis, that we falsify by abstraction.

Eddington has certainly been in part to blame for this, for he himself called his doctrines 'Selective Subjectivism', and introduced the two analogies which have dominated and confused later discussions. Suppose, he says, that an ichthyologist trawls the seas using a fish-net of two-inch mesh: then fish less than two inches in length will escape him, and he will find when he pulls up the net only fishes two inches long or more. This, Eddington suggests, may tempt him to conclude that the world contains no fish of smaller size; he may generalize and announce, "All fish are two inches long or more"; and until he has the sense to examine his own methods of fish-catching, he may fail to realize that these methods, not the ichthyological facts, are what have led him to the conclusion. This, Eddington argues, is what happens in physics: the theorist trawls the results of the experimenters' work through his net and announces as discoveries about the world things that he himself forces on the facts by his methods of trawling. Eddington also recalls the old story of Procrustes, the giant who obliged unfortunate travellers to sleep in his bed and always trimmed them to fit, stretching the shorter ones on the rack and lopping pieces off the longer ones until their corpses were exactly the right length. The theorist for him is Procrustes: the experimental observations are the travellers, and are adjusted willy-nilly until they fit exactly into the theoretical bed. "Let us therefore," Eddington implies, "be more self-conscious about our methods of theorizing, recognize that it is to subjectively selected data that the generalizations of physics—the so-called laws of nature —apply, and see what surprising things may not be discovered from a careful examination of our explanatory techniques."

One thing about Eddington's fish-net analogy must be pointed out at once. The conclusion which the incautious ichthyologist announces is one of natural history, an empirical generalization of the purest kind, "*All* fish share such-and-such a property". Elsewhere we have seen the disastrous effect the use of this model can have on our understanding of the physical sciences, and we must take care not to be misled here also. So let us pose Eddington's question in a way which is truer to life, and see how much of the problem remains. For these purposes the cartographical analogy is a useful guide, and can make Eddington's professional activities look less disreputable than they have tended to look to some of his colleagues.

We saw earlier how some features, even of the simplest theories, must be understood in terms of the method of representation we employ as much as of the phenomena represented. The central notion of geometrical optics, that of a light-ray, holds the centre of the theoretical stage only so long as the geometrical method of representation (ray-tracing) remains our basic technique of inference-drawing: as soon as the wave-theory displaces the simpler picture as the basic theory, the notion of a ray of light loses its theoretical importance. Nor is there anything mysterious about this, anything in particular which can be regarded as falsification of the facts. In cartography, too, there is a good deal which has to be contributed by us before there can be a map at all, and this contribution is again of an unmysterious kind. Cartographers and surveyors have to choose a base-line, orientation, scale, method of projection and system of signs, before they can even begin to map an area. They may make these choices in a variety of ways, and so produce maps of different types. But the fact that they make a choice of some kind does not imply in any way that they falsify their results. For the alternative to a map of which the method of projection, scale and so on were chosen in this way, is not a truer map—a map undistorted by abstraction: the only alternative is no map at all. To draw an analogy between a cartographer's method of projection and the ichthyologist's fish-net would accordingly be misleading. There is no question of falsification here. Quite the reverse: it is only after all these decisions have been taken and a map has been produced, that the question can even be raised, how far the product of the cartographer's work is true to the facts, for only then will there be anything which can be true to *or* falsify them.

If physicists are to be spoken of as in any way responsible for the structure of physical theory, the reasons are similar. For in physics, as much as in cartography, some decisions have to be taken, consciously or no, before a theory can be produced at all. If Eddington's remarks appear mysterious, this is probably because these decisions are so obvious, elementary and easily made that one is liable to overlook them, forget that they have ever been taken, and even take them without recognizing them for what they are. In geometrical optics, for instance, it is easy to forget that we have *decided* to represent optical phenomena by the use of lines drawn on paper or on the blackboard; and perhaps no one has come to understand the logic of physics who has not at some time been amazed that there should be *any* connexion between such things as shadows, lamps and patches of light on the one hand, and graphite streaks on paper on the other. The lines in our ray-diagrams are not, so to speak, thrown in with the phenomena: they have to be put into relation with the phenomena by our adoption of a particular theory, view of light and technique of representation. Wherever in physics we introduce numerical concepts, such as temperature, or employ mathematical techniques of inference of a geometrical or of a more sophisticated kind, decisions of this sort must have been taken.

Once again, this does not imply that the statements which the theoretical physicist advances for our acceptance are in fact falsehoods, which he is able to misrepresent as true as a result of his methods of theorizing. Here, too, the fish-net analogy is quite misleading. For the alternative to a theory which has been built up with the help of decisions of this kind is not a truer theory, 'free from the distorting effects of abstraction': the only alternative is no theory at all. Some contribution on our part to the structure of theoretical physics is needed if the statements within the theory are to be capable of having any application to the world; and only when this connexion has been established will there be anything to be spoken of either as 'true to the facts' or as 'falsifying the facts'.

The air of mystery and the suggestion of subjectivity, which have marked the discussion of Eddington's problem, are both therefore unnecessary. There is no need to feel that the physicist's contribution to his own theories is either personal, or necessarily unstateable: it is something as public, and as open to inspection and description, as a cartographer's methods of projection and representation. Reading about this subject, as when reading Kant, one gets the impression that to try to say where to draw the line between our own contribution and that of the facts is in some curious way an impossibility—rather like trying to chew your own teeth. But this is a mistake. It is not that the physicist has a mysterious predilection for some theoretical mould, into which he thrusts all the experimental results he meets, nor is it a deep necessity of experience that he should handle these results in the way he does. His part is no more than that played by anyone who introduces a language, symbolism, method of representation or system of signs.

Perhaps if Kant's arguments were stripped of their unhappy air of psychological discovery and re-expressed in similar terms, they too would cease to be so obscure. For if the decisions on which our physical theories rest are easy to forget, those which have gone to the making of everyday speech are yet more easily forgotten; and the philosophical effects of forgetting them, as Wittgenstein saw, are yet more pervasive. To talk, in the philosophy of science, of theoretical physics falsifying by abstraction, and to ask for the facts and nothing but the facts, is to demand the impossible, like asking for a map drawn to no particular projection and having no particular scale. In epistemology, too, to argue that our everyday concepts falsify by abstraction or are necessary conditions of experience, with the suggestion that one thereby points to a defect in our conceptual equipment or to an unfortunate limitation on our capacity for experiencing, is to evince a similar misconception. If we are to say anything, we must be prepared to abide by the rules and conventions that govern the terms in which we speak: to adopt these is no submission, nor are they shackles. Only if we are so prepared can we hope to say anything true—or anything untrue. It is unreasonable to complain, as philosophers have so often done, because we cannot tell the truth without talking.

4.6 *Facts and Concepts: the Absolute Zero*

In order to indicate what sort of thing the physicist's contribution to his theories consists of, let us look at a simple example. For it is possible to show, with a very little technical explanation, how the acceptability of statements which at first glance seem to be pure matters of fact may depend, rather, on the technique of representation employed in a physical theory.

A suitable example is at hand if one considers the physicist's notion of temperature. When one first learns about temperature and about thermal phenomena, the existence of the Absolute Zero of temperature may appear to one as a strange and ineluctable fact about the Universe. The world of thermal phenomena, it seems, has a curious and unforeseen feature. As we work our way down lower and lower, we cannot go on for ever, but after a time come up against an adamantine layer, against which even our best drills are blunted: all attempts to penetrate it are in vain. The existence of the Absolute Zero may thus present itself to us as the brutest of brute facts; and the natural geological analogy, between up and down in temperature and up and down from ground-level, reinforces this impression. Of course the Absolute Zero is not something which one comes up against with a bang: rather, as one produces lower and lower temperatures, all further reductions get harder to make, so that at $-270°$C. it may be more difficult to cool things by $\frac{1}{10}°$C. than, at ordinary temperatures, it is to cool them by $10°$C. But the geological picture will accommodate this additional feature easily enough: it is as though, as our drills went down, we came up against progressively more impenetrable strata, the Absolute Zero being the limit beyond which, it seems, there will never be any hope of piercing, however much we improve our drills.

This geological picture is totally misleading. The existence, at some point, of an Absolute Zero of temperature is not a brute fact at all, but a conceptual matter—i.e. a consequence of the way in which we give a meaning to the notion of temperature, and put degrees of warmth and cold into relation with the number-series. We who grow up familiar with thermometers tend to overlook the fact that this has had to be done. Yet there is no more connexion between numbers and the notions of heat and cold, until we create one, than there is between pencil-marks on paper and optical phenomena. In either case, some-one had the genius to see what a help it would be to introduce a new concept ('light-ray' or 'temperature'), and so the crucial steps were taken. When Galileo invented the notion of temperature and designed the first thermometer, he knew very well what he was doing. He saw that to produce a thermometer would not just be to find a way of measuring something which before we had been able to estimate only roughly: rather, it would be to alter the whole status of our thermal notions. He did what he did as part of a deliberate campaign, the first stage in his programme of making physics mathematical, and 'turning secondary qualities into primary ones'. Likewise, the physicists who helped to extend our scale of temperature were not just developing fresh instrumental techniques, but helping to fix the meaning of the term 'temperature' in respects in which it had previously been indeterminate. This shows why the title 'theory of measurement', which has often been used for our present field of discussion, may be misleading. Techniques of measurement and conceptual refinements do often proceed *pari passu*, but for logical purposes we must keep conceptual matters distinct from questions of experimental technique.

If one wants to understand about the Absolute Zero, the crucial thing to examine is the introduction of the ideal gas scale of temperature as the basic theoretical scale. This scale is introduced by three steps. First, it is remarked that the behaviour of all gases tends to conform the more nearly to a single law, the more we heat them up and the lower we make their pressures. This law is Charles' Law, according to which each degree through which we heat or cool a closed container of gas, as measured, for instance, on a mercury thermometer, should produce the same change of pressure whatever the gas. The more we cool different gases down, on the other hand, and the more we increase their pressures, the more markedly their behaviours diverge from each other: they liquefy and solidify at quite different temperatures from one another, and their compressibilities vary more and more as they approach the temperature of condensation.

Next, the common behaviour of all gases at high temperatures and low pressures is taken as a theoretical standard, deviations from which require to be explained. To mark the adoption of this standard, physicists proceed to introduce the notion of an ideal gas, which is defined as one behaving at all temperatures in the manner in which actual gases tend the more nearly to behave, the higher the temperature and the lower the pressure. This notion is, of course, even more of a theoretical ideal than that of a light-ray. Finally, temperature on the ideal gas scale is introduced by reference to the properties of this ideal gas: equal changes in temperature, on this scale, are defined as those which produce equal changes of pressure in a closed container of ideal gas. To measure temperature on this scale thermometers containing simple gases, such as hydrogen, are used, their readings being corrected where necessary to allow for deviations from the theoretical scale.

Now notice one thing about the ideal gas scale: it cannot help having an Absolute Zero. For, whatever may be the pressure of a given mass of ideal gas when it occupies one cubic centimetre at the freezing point of water, it will not make sense to talk of cooling it down by more degrees of temperature below 0°C. than will reduce this pressure to zero. The precise numerical value of the Absolute Zero, in degrees Centigrade, is a brute fact which has to be found out by investigating the properties of actual gases at high temperatures. But that there is an Absolute Zero *at all* is something which does not have to be found out by experiment, being ensured by our way of introducing the ideal gas scale. It turns out in fact to be —273.16°C. This figure was, of course, known very precisely long before physicists had any means of approaching it in practice. It is a conceptual matter, a fact about our *notion* of temperature, not as one might at first suppose, a fact about thermal phenomena at very low temperatures.

The statement, "Nothing can be cooled below the Absolute Zero" or, to put the same thing less misleadingly, "The ideal gas scale has a lower bound", is accordingly one of those theoretical statements which may look at first like a fact about actual phenomena; but which turns out on closer inspection to be a consequence of the technique of representation adopted— in this case, of the particular manner in which the notion of temperature is fitted into our theories. The existence of the Absolute Zero can be compared with the existence of the boundary in a map of the World drawn to a stereographic or orthographic projection. On these projections, the surface of the Earth does not cover the whole of any sheet of paper you use, as a Mercator's map is capable of doing, but fills only two circles. If there is blank space round the circles, that is not because the cartographer has chosen to cut off the map half-way up Greenland, say, but because, the nature of the projection being what it is, no point on the Earth can be mapped outside the circles. One can, of course, decide to make the circles as large as one chooses; but, however large one decides to have them, there will still be a boundary, whereas a map drawn to Mercator's projection is capable of going on indefinitely.

If we prefer, it is open to us to stop using a map of one kind and start using one of the other kind; and to abolish the boundary in this way shows nothing about the area we are mapping. The presence or absence of such a boundary tells us nothing about the surface of the Earth. The same is true in physics. One can, if one chooses, change over from the ordinary ideal gas scale to a logarithmic scale, which extends without limit in both directions; and to make this change implies nothing about actual thermal phenomena. In neither case does one, by changing the method of representation, burke any facts about the World.

Here the defects of the geological analogy become clear. For so long as we think in terms of this picture, the inaccessible strata below the adamantine layer seem as authentic as those above it: that is why it seems a simple question of fact that we cannot break through to the 'inaccessible' temperatures below the Absolute Zero. But the truth is quite otherwise. The way we line up degrees of warmth and cold with numbers in the ideal gas scale is such that numbers below -273.16 are given no interpretation as temperatures: all the thermal phenomena that are conceived of in the current theories are mapped on to the range of numbers from -273.16 upwards. So the inaccessible temperatures below the Absolute Zero are a myth. On our standard theoretical scale, figures like '-300' no more represent inaccessible temperatures than do the blank spaces round a stereographic map represent inaccessible places: all genuinely inaccessible places, such as the top of Mt. Everest, have a place within the circles, quite as much as Leicester Square. It is true that our theories may perhaps come to be altered some day, and a fresh temperature-scale introduced along with new theories, but there is no reason to anticipate this; and in any case, if it happens, it will not mean that a new, sharper drill has been built which has torn a way through the adamantine layer, but rather that we, who put the layer there to begin with, have moved it elsewhere.

4.7 *Do sub-microscopic entities exist?*

Non-scientists are often puzzled to know whether the electrons, genes and other entities scientists talk about are to be thought of as really existing or not. Scientists themselves also have some difficulty in saying exactly where they stand on this issue. Some are inclined to insist that all these things are just as real, and exist in the same sense as tables and chairs and omnibuses. But others feel a certain embarrassment about them, and hesitate to go so far; they notice the differences between establishing the existence of electrons from a study of electrical phenomena, inferring the existence of savages from depressions in the sand, and inferring the existence of an inflamed appendix from a patient's signs and symptoms; and it may even occur to them that to talk about an electromagnet in terms of 'electrons' is a bit like talking of Pyrexia of Unknown Origin when the patient has an unaccountable temperature. Yet the theory of electrons does *explain* electrical phenomena in a way in which no mere translation into jargon, like 'pyrexia', can explain a sick man's temperature; and how, we may ask, could the electron theory work at all if, after all, electrons did not really exist?

Stated in this way, the problem is confused: let us therefore scrutinize the question itself a little more carefully. For when we compare Robinson Crusoe's discovery with the physicist's one, it is not only the sorts of discovery which are different in the two cases. To talk of existence in both cases involves quite as much of a shift, and by passing too swiftly from one use of the word to the other we may make the problem unnecessarily hard for ourselves.

Notice, therefore, what different ideas we may have in mind when we talk about things 'existing'. If we ask whether dodos exist or not, i.e. whether there are any dodos left nowadays, we are asking whether the species has survived or is extinct. But when we ask whether electrons exist or not, we certainly do not have in mind the possibility that they may have become extinct: in whatever sense we ask this question, it is not one in which 'exists' is opposed to 'does not exist any more'. Again, if we ask whether Ruritania exists, i.e. whether there is such a country as

Ruritania, we are asking whether there really is such a country as Ruritania or whether it is an imaginary, and so a non-existent country. But we are not interested in asking of electrons whether they are genuine instances of a familiar sort of thing or non-existent ones: the way in which we are using the term 'exist' is not one in which it is opposed to 'are non-existent'. In each case, the word 'exist' is used to make a slightly different point, and to mark a slightly different distinction. As one moves from Man Friday to dodos, and on from them to Ruritania, and again to electrons, the change in the nature of the cases brings other changes with it: notably in the way one has to understand sentences containing the word 'exist'.

What, then, of the question, "Do electrons exist?" How is this to be understood? A more revealing analogy than dodos or Ruritania is to be found in the question, "Do contours exist?" A child who had read that the equator was 'an imaginary line drawn round the centre of the earth' might be struck by the contours, parallels of latitude and the rest, which appear on maps along with the towns, mountains and rivers, and ask of them whether *they* existed. How should we reply? If he asked his question in the bare words, "Do contours exist?", one could hardly answer him immediately: clearly the only answer one can give to this question is "Yes and No." They 'exist' all right, but do they *exist*? It all depends on your manner of speaking. So he might be persuaded to restate his question, asking now, "Is there really a line on the ground whose height is constant?"; and again the answer would have to be "Yes and No", for there is (so to say) a 'line', but then again not what you might call a *line*. . . . And so the cross-purposes would continue until it was made clear that the real question was: "Is there anything to show for contours—anything visible on the terrain, like the white lines on a tennis court? Or are they only cartographical devices, having no geographical counterparts?" Only then would the question be posed in anything like an unambiguous manner. The sense of 'exists' in which a child might naturally ask whether contours existed is accordingly one in which 'exists' is opposed not to 'does not exist any more' or to 'is non-existent', but to 'is only a (cartographical) fiction'.

This is very much the sense in which the term 'exists' is used of atoms, genes, electrons, fields and other theoretical entities in the physical sciences. There, too, the question "Do they exist?" has in practice the force of "Is there anything to show for them, or are they only theoretical fictions?" To a working physicist, the question "Do neutrinos exist?" acts as an invitation to 'produce a neutrino', preferably by making it *visible*. If one could do this one would indeed have something to show for the term 'neutrino', and the difficulty of doing it is what explains the peculiar difficulty of the problem. For the problem arises acutely only when we start asking about the existence of *sub-microscopic* entities, i.e. things which by all normal standards are invisible. In the nature of the case, to produce a neutrino must be a more sophisticated business than producing a dodo or a nine-foot man. Our problem is accordingly complicated by the need to decide what is to count as 'producing' a neutrino, a field or a gene. It is not obvious what sorts of thing ought to count: certain things are, however, generally regarded by scientists as acceptable—for instance, cloud-chamber pictures of α-ray tracks, electron microscope photographs or, as a second-best, audible clicks from a Geiger counter. They would regard such striking demonstrations as these as sufficiently like being shown a live dodo on the lawn to qualify as evidence of the existence of the entities concerned. And certainly, if we reject these as insufficient, it is hard to see what more we can reasonably ask for: if the term 'exists' is to have any application to such things, must not this be it?

What if no such demonstration were possible? If one could not show, visibly, that neutrinos existed, would that necessarily be the end of them? Not at all; and it is worth noticing what happens when a demonstration of the preferred type is not possible, for then the difference between talking about the existence of electrons or genes, and talking about the existence of dodos, unicorns or nine-foot men becomes all-important. If, for instance, I talk plausibly about unicorns or nine-foot men and have nothing to show for them, so that I am utterly unable to say, when challenged, under what circumstances a specimen might be, or might have been seen, the conclusion may reasonably be drawn that my nine-foot men are imaginary and my unicorns a myth. In either case, the things I am talking about may be presumed to be non-existent, i.e. are discredited and can be written off. But in the case of atoms, genes and the like, things are different: the failure to bring about or describe circumstances in which one might point and say, "There's one!", need not, as with unicorns, be taken as discrediting them.

Not all those theoretical entities which cannot be shown to exist need be held to be non-existent: there is for them a middle way. Certainly we should hesitate to assert that any theoretical entity really existed until a photograph or other demonstration had been given. But, even if we had reason to believe that no such demonstration ever could be given, it would be too much to conclude that the entity was non-existent; for this conclusion would give the impression of discrediting something that, as a fertile explanatory concept, did not necessarily deserve to be discredited. To do so would be like refusing to take any notice of contour lines because there were no visible marks corresponding to them for us to point to on the ground. The conclusion that the notion must be dropped would be justified only if, like 'phlogiston', 'caloric fluid' and the 'ether', it had also lost all explanatory fertility. No doubt scientists would be happy if they could refer in their explanations only to entities which could be shown to exist, but at many stages in the development of science it would have been crippling to have insisted on this condition too rigorously. A scientific theory is often accepted and in circulation for a long time, and may have to advance for quite a long way, before the question of the real existence of the entities appearing in it can even be posed.

The history of science provides one particularly striking example of this. The whole of theoretical physics and chemistry in the nineteenth century was developed round the notions of atoms and molecules: both the kinetic theory of matter, whose contribution to physics was spectacular, and the theory of chemical combinations and reactions, which turned chemistry into an exact science, made use of these notions, and could hardly have been expounded except in terms of them. Yet not until 1905 was it definitively shown by Einstein that the phenomenon of Brownian motion could be regarded as a demonstration that atoms and molecules really existed. Until that time, no such demonstration had ever been recognized, and even a Nobel prize-winner like Ostwald, for whose work as a chemist the concepts 'atom' and 'molecule' must have been indispensable, could be sceptical until then about the reality of atoms. Moreover by 1905 the atomic theory had ceased to be the last word in physics: some of its foundations were being severely attacked, and the work of Niels Bohr and J. J. Thomson was beginning to alter the physicist's whole picture of the constitution of matter. So, paradoxically, one finds that the major triumphs of the atomic theory were achieved at a time when even the greatest scientists could regard the idea of atoms as hardly more than a useful fiction, and that atoms were definitely shown to exist only at a time when the classical atomic theory was beginning to lose its position as the basic picture of the constitution of matter.

Evidently, then, it is a mistake to put questions about the reality or existence of theoretical entities too much in the centre of the picture. In accepting a theory scientists need not, to begin with, answer these questions either way: certainly they do not, as Kneale suggests, commit themselves thereby to a belief in the existence of all the things in terms of which the theory is expressed. To suppose this is a variant of the Man Friday fallacy. In fact, the question whether the entities spoken of in a theory exist or not is one to which we may not even be able to give a meaning until the theory has some accepted position. The situation is rather like that we encountered earlier in connexion with the notion of light travelling. It may seem natural to suppose that a physicist who talks of light as travelling must make some assumptions about what it is that is travelling: on investigation, however, this turns out not to be so, for the question, what it is that is travelling, is one which cannot even be asked without going beyond the phenomena which the notion is originally used to explain. Likewise, when a scientist adopts a new theory, in which novel concepts are introduced (waves, electrons or genes), it may seem natural to suppose that he is committed to a belief in the existence of the things in terms of which his explanations are expressed. But again, the question whether genes, say, really exist takes us beyond the original phenomena explained in terms of 'genes'. To the scientist, the real existence of his theoretical entities is contrasted with their being only useful theoretical fictions: the fact of an initial explanatory success may therefore leave the question of existence open.

There is a converse to this form of the Man Friday fallacy. Having noticed that a theory may be accepted long before visual demonstrations can be produced of the existence of the entities involved, we may be tempted to conclude that such things as cloud-chamber photographs are rather overrated: in fact, that they only seem to bring us nearer to the things of which the physicist speaks as a result of mere illusion. This is a conclusion which Kneale has advanced, on the ground that physical theories do not stand or fall by the results obtained from cloud-chambers and the like rather than by the results of any other physical experiments. But this is still to confuse two different questions, which may be totally independent: the question of the acceptability of the theories and the question of the reality of the theoretical entities. To regard cloud-chamber photographs as showing us that electrons and α-particles really exist need not mean giving the cloud-chamber a preferential status among our grounds for accepting current theories of atomic structure. These theories were developed and accepted before the cloud-chamber was, or indeed could have been invented. Nevertheless, it was the cloud-chamber which first showed in a really striking manner just how far nuclei, electrons, α-particles and the rest could safely be thought of as real things; that is to say, as more than explanatory fictions.

CHAPTER V

UNIFORMITY AND DETERMINISM

IT is often said in philosophical discussions about the sciences that either they, or scientists, or scientific arguments presuppose (or take for granted, or assume) some fact (or general principle, or major premise) which is spoken of as 'the Uniformity of Nature'. It is time to re-examine this notion, and see what light is thrown on it by the results of our discussion.

This extremely vague way of introducing the subject has been chosen deliberately. Different writers present the Uniformity of Nature in different guises, feel bound to invoke the idea for different reasons, and formulate it very differently. Some see in it the solution of the problem of induction, the logical bridge spanning the 'gulf' between observations made in the past and predictions made about the future; others see it as an article of scientific faith, the expression of the scientist's confidence in the possibility of solving his problems; others again look backwards at the achievements of science in past centuries, and see in them evidence of uniformity already revealed. The arguments advanced and the points made differ correspondingly, and each requires a separate examination. All that we can hope to do in this chapter is to put the doctrine in a form which bears directly on the physical sciences, and see what light is thrown on it by our examination of the types of arguments physicists have occasion to employ.

5.1 *Are laws of nature universally applicable?*

The Principle of the Uniformity of Nature, then, takes many forms, and is asked to do many jobs. Let us begin by considering one of the more extreme suggestions in connexion with which it appears: namely, the doctrine that, taken by themselves, the arguments employed in the physical sciences are logically unsound, and that the holes in them can be plugged only by introducing as a major premise in all such arguments some statement about the uniformity of things-in-general. Have philosophers good reasons for thinking that any such extra premise is called for? Is it in fact needed? And, if anything is, in some sense or other, assumed in scientific arguments, will a general principle about uniformity help to justify the assumptions made? From Mill's *System of Logic* down to Russell's *Human Knowledge*, a long and important series of writers has answered "Yes" to all these questions.

One thing springs to the eye as soon as one begins to watch scientists at work, which strongly suggests that assumptions are being made. When, for instance, physicists calculate the manner in which falling apples, the moon, the satellites of Jupiter and double-stars many light-years away may each be expected to move, they employ the same law of gravitation in each case: it does not seem to occur to them that the form of the equation might need to be modified in passing from one system to the next. Yet surely it should: surely, one may feel, it is a question whether the law of gravitation is the same on Mars —and *a fortiori* in the distant nebulae—and surely a million years ago its form might not have been the same. Does not this fact alone show that pre-suppositions are being made about the Uniformity of Nature? To be specific: is it not being assumed that the laws of nature take, will take and have always taken one form everywhere—the same in distant parts of the Universe and at remote epochs as here and now?

Certainly if laws of nature are put in the same pigeon-hole as empirical generalizations, this conclusion seems irresistible; and it is easy to see how, having this analogy in mind, one might come to accept it without question. Suppose, for instance, that there turned out to be cats and rabbits on Mars as well as on the Earth. Then it is quite on the cards that their diets would be completely unlike those of their terrestrial fellows. It might be the case, for instance, that the rabbits on Mars ate nothing but mice, while Martian cats turned up their noses at flesh and lived on lettuces. At any rate, the generalizations that cats are carnivores and that rabbits live on lettuces—both of which are what we have called habit-statements—will require further checking as soon as the Martian species are discovered; and it will be dangerous for natural historians to jump to conclusions. To say, e.g. "So there are rabbits on Mars, are there? Well, then, there must be vegetables there, too, for them to live on," would very definitely be to take it for granted that the habits of rabbits were the same there as here. Granted that this is so in the case of rabbits, what about gravity? Might not gravity too work differently on Mars, and the law of gravitation itself be different there from here? Are not physicists running as much of a risk in taking the answer for granted as natural historians would be if they assumed too much about Martian rabbits?

To get the first indication of the right answer to this question, compare for a moment, not the things which are allegedly assumed to be the case, but those things which are allegedly assumed *not* to be the case. This is worth doing because, for an assumption to amount to anything, it must rule out some possibility: an assumption which could be invalidated by no describable happening is only misleadingly so called. Now the two assumptions are, respectively, that rabbits eat the same sorts of food everywhere, and that the law of gravitation takes the same form everywhere. So what is said to be ruled out is, in the one case, that rabbits eat elsewhere kinds of food they do not eat on the Earth; and in the other, that the law of gravitation takes a different form elsewhere from that which it takes on the Earth. The first of these two possibilities is, clearly, a perfectly meaningful one: one knows well enough what it would be like to discover rabbits eating mice rather than lettuces. But what about the second one? This requires a closer examination; first there is an ambiguity to be resolved; and when this has been done it will be questionable whether after all, in the required sense, the suggestion means anything.

To deal with the ambiguity first—what are we to understand by the phrase 'the law of gravitation'? As in Chapter III, we must distinguish between Newton's inverse-square law or Einstein's refinement of it, these being the sorts of thing one might call 'laws of nature'; and such a statement as that freely falling bodies accelerate by 32.2 ft./sec. every second, which might loosely be called 'the law of gravitational acceleration on the Earth', but is not the sort of thing physicists would call a law of nature. What we have to say about the alleged assumption depends entirely on which of these we consider. It would certainly be dangerous to assume that freely falling bodies accelerated at 32.2 ft./sec./sec. on Mars, as they do on the Earth. But this is not an assumption of a sort that physicists would dream of making. The rate of acceleration of bodies moving under gravitational attraction alone, expressed in ft./sec./sec., and other such specific gravitational effects, will without doubt be different on Mars from what they are on the Earth: in this sense, gravity unquestionably works differently, and the law of gravitational acceleration will be different, on Mars. In this sense, the assumption that physical laws operate in the same way everywhere may make perfectly good sense; but it is also quite unfounded, and physicists would never make it. This example, however, goes no way towards establishing the need for a general Principle of the Uniformity of Laws of Nature, since the so-called 'law' considered is not a law of nature at all, but rather an empirical discovery which is to be accounted for by applying the law of gravitation to the special circumstances of the Earth.

If we turn to a real law of nature, the situation changes at once, since there is now no room to say what it is that is being ruled out. For if, as one might conceive happening, the study of gravitational phenomena on Mars obliged us to amend the law of gravitation, we could not let things rest there: we could not cheerfully say "The law takes a different form on Mars", as we might say "Rabbits eat different food on Mars". Our conception of a law of nature requires that, if the law has to be so amended, the modified law must continue to explain the terrestrial and other phenomena previously accounted for in terms of the unmodified law. Any discovery which forced us to amend the law of gravitation itself would therefore be regarded as revealing not a gravitational non-uniformity, but rather an inadequacy in our present ideas about gravity, a defect in the theory of gravitation with implications as much for the Earth and the distant nebulae as for Mars. If there were, say, uncommonly strong gravitational fields on Mars, such an inadequacy might be shown up first by a study of Martian gravity; but once discovered it could not be put on one side with a mere "They order these things differently on Mars".

The statement that the law of gravitation might be different on Mars is, therefore, of doubtful meaning. This being so, the suggested assumption that it is not the case that the law is different on Mars is also of doubtful meaning. So after all, it is not as clear as it at first seemed to be that physicists are assuming anything when they apply the same law of gravitation to gravitational phenomena in different places or at different times.

5.2 *Physicists work on presumptions, not assumptions*

The point is worth expanding. All we have shown so far is this: that, by itself, a physicist's expressing his Law of Gravitation in an identical form on all occasions proves nothing about the Uniformity of Nature. To add the conclusion, that employing the same law to account for all gravitational phenomena does not entail making any assumptions, may be misleading. For what are we to count as gravitational phenomena? Once we have a theory of gravitation of any standing, it will be just those phenomena which can be, or which there is reason to suppose might be explained in terms of that theory that will be called 'gravitational' phenomena, so our conclusion is at first sight an empty one.

But while the conclusion may rest at bottom on a tautology, it is none the less an important one. Once again, the thing that matters is the difference between laws and generalizations. In physics, if we start by taking a phenomenon to be purely gravitational, but it turns out not to be properly explicable on the current theories, then one of two things may happen: either we must conclude that the phenomenon is not after all a purely gravitational one, and look elsewhere for an explanation of the deviations from the expected behaviour; or we must call the current theory of gravitation in question. In the former case, other laws must be brought in to account for the unforeseen features of the phenomenon: in the latter, the laws and theories must be reconsidered and revised. But if reconsideration is forced on us, the revised law (e.g. Einstein's) will no more be expressed in a way referring to a particular place or occasion than was the unmodified law (e.g. Newton's): it will accordingly be applicable, if at all, to every appropriate system of bodies, regardless of the place and time in question. In neither case will there be any room to talk of a law of nature having a different form at different places or times; nor, except misleadingly, of its having the *same* form at different places and times either, for only one expression can be entitled to such a name as 'the Law of Gravitation'.

Generalizations, such as we meet in natural history, are treated in quite a different way. The statement "All rabbits eat lettuces" is liable to sudden upset, and might well be rewritten "All known species of rabbit eat lettuces". Here there is an unspoken reservation—"On Mars perhaps, who knows?" The generalization requires to be covered by a guarding clause of a kind which would be totally out of place if added to a law of nature.

This difference between laws and generalizations is connected with something we noticed earlier, the fact that natural historians are committed for the most part to the everyday classification of their subject-matter, whereas it is open to physical scientists to reclassify theirs as they go along: "What is or is not a cow is for the public to decide"—but how different it is with cadmium, a diffraction pattern, an electron or a meson field. Each term in the generalization "Rabbits eat lettuces" is accordingly given a meaning before the generalization is, or indeed could be formulated. Supposing a zoologist were faced with lettuce-loathing rabbits brought from Mars by interplanetary travellers, it would not be open to him to say, "Let us call these 'tibbars', to distinguish them from rabbits which, as everyone knows, eat lettuce". He might, if he chose, give the Martian rabbits a different Latin name and a special taxonomical status, but he would not be entitled to resort to *ad hoc* reclassification simply to save the everyday generalization that "Rabbits eat lettuces". For him to do so would be like a man's saying "No Briton would lay violent hands on a woman", and then trying to save his claim from falsification in the light of the Law Reports by amendments of the form "No *true* Briton would . . ."—a patriotic move, perhaps, but a logically un-systematic one, of a kind for which there can be no room in the sciences.

In formalized sciences such as physics, by contrast, the terminology is not fixed beforehand, least of all by the public. Theories, techniques of representation and terminologies are introduced together, at one swoop. It is thereafter a technical question, in what specifiable circumstances a given metallic strip can be accepted as cadmium, some apparatus regarded as a neutron source, or a particular body treated for theoretical purposes as moving under gravitational influences alone. The application of a chemical substance-name such as 'cadmium' involves much more than the use of words like 'wooden' or 'stone', for it not only labels the specimen by origin and every-day characteristics, but places it on the physico-chemical map. We shall have to look more carefully at this point later. It is the same with a phrase like 'purely gravitational phenomenon': to use this phrase for the motion of some system such as a double-star is likewise to place it on the physical map—to commit oneself to the belief that the phenomenon can be explained by the application of some one particular theory alone. The phenomena which form the physicist's field of study are classified in a systematic way, which reflects the terms in which and the methods by which he sets about explaining them; and it is the systematic nature of this reclassification which dis-tinguishes it from *ad hoc* unsystematic distortions of an existing classification, like the "No true Briton" move.

The fact that physicists always speak of one and the same thing as their 'law of gravitation', regardless of the place and time referred to, involves them, in consequence, in no par-ticular assumptions: it would not be a law of nature if they did otherwise. They would be making assumptions only if, e.g., they were to suppose that all the systems they studied would turn out to be purely gravitational, and so did not bother to consider the possibility that other kinds of theory besides gravitation theory would be required to account for their be-haviour. But this sort of assumption, again, is one they would never make. They never assume that all the systems they study are of one type: the most that they do is to *presume* (a) that the existing theories will, between them, suffice to explain the behaviour of each fresh system of bodies which they choose to study and (b) that any fresh system of bodies they examine will resemble most closely in behaviour those systems which it most closely resembles in structure. This can be illustrated with the help of our original example. The fact that an astrophysicist uses the same law, when explaining the motion of the parts of a double-star, as has already been used to account for the motion of falling apples, the moon and the satellites of Jupiter, represents a uniformity in his techniques for dealing with the four systems. This uniformity in technique reflects the presumption that the four phenomena can be regarded as similar in type, viz. gravitational: it would not be found if we looked at physicists working in different fields—gravitation and magnetism, say.

Further, a physicist's presumptions are only *initial* presumptions. Our astrophysicist, for instance, must be on the look-out for deviations; and if he finds things working out otherwise than he was led by gravitation theory to expect, he will have to ask himself why this happens. If he were not on the look-out for such deviations, and did not even bother to ask whether the theory of gravitation would explain the star's motions by itself, or whether other forces were involved, then one might indeed say that he was presupposing or taking for granted something which was in need of justification. But, in fact, he will always be ready to reconsider the initial presumption that a double-star and the solar system are, theoretically speaking, strictly comparable—as soon as there is any reason to do so. As soon as he does begin to look elsewhere for an explanation of observed deviations, he will have abandoned the initial presumption that the motion of the parts of the double-star can be regarded as purely gravitational: presumption (b) above. But this is not all. If and when there is an adequate reason for doing so, he will abandon also the deeper presumption (a), that there is a place for the novel system on the map as it is, i.e. that the existing theories between them are capable of explaining the behaviour of the new double-star, and

will try to discover how the existing theories can be modified
or supplemented in order to account for its behaviour.

What to begin with looked like an assumption turns out
therefore to be hardly more than a piece of common sense. If
physicists use the same form of law in widely differing cases,
that is the mark, not of a daring presupposition about the
Uniformity of Nature, but of a decently methodical procedure.
And if we try to express in words what it is that physicists
thereby presume, it will take the form, not of a grandiose
principle about things-in-general, but rather of some such trite
expression as this: that, unless there is some reason to suppose
that a novel phenomenon cannot be explained in terms of the
theory which it is natural to turn to first, there is every reason
to turn first to that theory. This is not a very dangerous presumption, in need of a reasoned defence. In any case, it is one
which carries its own shield with it: if in practice a physicist is
mistaken in his first attempts at an explanation, taking for a
purely magnetic phenomenon, say, what on further investigation turns out to be partly an electrical one, this is something
which will soon show up. Once he has found out his mistake,
he will be warned, and will know what to expect next time he
encounters a similar system. So it is not Nature that is Uniform,
but scientific procedure; and it is uniform only in this, that it is
methodical and self-correcting.

5.3 *Criteria of similarity within and outside science*

One last attempt might be made to 'make honest generalizations' out of the statements of theoretical physics, and so to
find a place in the arguments of physical theory for a Principle of
the Uniformity of Nature. For on the previous page we spoke of
systems being 'similar in structure': might one not accordingly
say, at the very least, that physicists assume similar phenomena
to occur always when structurally similar systems are placed in
similar situations?

This suggestion is an attractive one only so long as we leave
the physicists' criteria of similarity unexamined. For where do
these criteria come from? Suppose that phenomena and
situations were to be classified as similar within the physical

sciences on the same grounds as outside: if this were so, then we might yet have material for the kind of overriding generalization which could be spoken of as a 'Principle of the Uniformity of Nature', without being a complete truism. But is it so?

At first sight the everyday criteria of similarity, with which we are familiar outside physics, seem to fit well enough. One might reasonably claim to see a resemblance between a punt-pole sticking out of the river and a walking-stick half-immersed in a water-butt; and one might suppose that this resemblance accounted for the similarity in the explanations which a physicist would give of the way the two things looked. So, one might conclude, the resemblance between physically similar situations is something which can be seen; and the Uniformity Principle can be put in the form, "Where two systems can be seen to resemble one another, the explanations of their behaviour are similar." But one has only to look at a punt-pole sticking out of the river and another punt-pole lying broken on the river-bank in order to see a resemblance between them, too, yet in this case the explanations a physicist would give of the way they looked would be quite unlike one another: whatever resemblance we may see between the objects is, for physical purposes, irrelevant. This kind of thing happens very often. Even where there are resemblances to be seen, these may be of no interest to the physicist. The reason is that the criteria of physical similarity between phenomena, objects and situations are fixed by our experience within physics, and not beforehand. So the statement, "Similar phenomena occur always when similar systems are placed in similar situations", is true only if one counts as similarities those resemblances, and those alone, which turn out to be physically significant; and then it is not so much a generalization as a truism.

To know what phenomena, systems or situations to speak of within physics as 'similar' requires not merely an eye for resemblances, but a knowledge of what resemblances matter; and this knowledge comes only when one has some acquaintance, however rudimentary, with the theories physicists have come to accept. If you point out as similar phenomena the ways in which a punt-pole looks in the river and a walking-stick looks in a water-butt, you thereby show your familiarity with elementary optics: had you no such familiarity, you could not know that the seen resemblance was physically significant, i.e. that the explanations would be sufficiently alike to justify your remark. Consider a contrasted example: if your wireless howls and the man comes from the radio shop to have a look at it, he may invoke all sorts of resemblances in the course of his diagnosis which are far from obvious to a layman. Perhaps the aerial is festooned around the valves, and the mechanic points to it, saying, "If you do that sort of thing with your set, you can't expect good reception: why, it's like trying to hold a public meeting in the Dome of St. Paul's." No doubt you will find this remark mystifying: the resemblance between your coiled aerial and the Dome of St. Paul's may not be striking. Yet there is nothing fundamentally different about the example: only here it is manifest that the criteria of physical likeness depend entirely on the formulation of a satisfactory theory. No doubt the mechanic will use some technical term such as 'resonance' to mark the likeness; but this acts simply as a sign-post pointing towards the theory which justifies the comparison. This function of technical terms seems to be overlooked by Mach, who tends to use words like 'refraction' and 'diffraction' in a misleading way; as though the layman could tell diffraction from refraction as surely as he can tell a cow from a pig. Whereas it is only with the development of optical theories that the need for fresh terms and fresh criteria of similarity becomes clear.

If we try to formulate the Principle of the Uniformity of Nature in terms of similarities between different phenomena and different situations, the result will be either vacuous or untrue. One can say, of any particular phenomenon, "Things always happen in that way under such circumstances", and to do so shows that you have a grasp of the particular factors required if an explanation of this phenomenon is to be given. But that is all; and there is no place for any general statement, any common-sense generalization, which will in all cases say what factors are physically significant. This is something which has to be discovered afresh in each part of the subject.

There seems, then, no hope of finding a place for the proposed Principle of Uniformity as a *premise* in the arguments of physicists. This conclusion is borne out if we look at an argument in which one might genuinely talk of a premise being assumed. For instance, there has been developed recently a method of dating archaeological finds, known as the 'radiocarbon method'. Wherever a find includes organic remains— bones of animals or men, or ashes, or relics of wooden structures or implements—the date of death can be computed from the amount of radio-active carbon present in the remains. Now this calculation can be made only on the assumption that, during the lifetime of the animals or men or trees concerned, the radiocarbon content of the atmosphere was effectively the same as it is now; for it is the decrease since that time in the proportion of radio-carbon present in the remains from which is computed the lapse of time. Here we have a very genuine assumption, and one which in fact there is every reason to suppose reliable. If, however, any reason were shown for modifying the assumption —if, for instance, evidence were found that the radio-carbon content of the atmosphere had been greater 10,000 years ago than it is now—then all our calculations would have to be reviewed, and the dates inferred from radio-carbon measurements would have to be altered. This is the mark of a genuine assumption: modify the assumption, and the conclusions will change.

What would change if we gave up the Principle of the Uniformity of Nature? How would it alter our scientific conclusions if we modified this assumption? This is never explained; and it is not easy to see how scientists could be led, without it, to conclusions other than those they reach anyhow. This being so, it is better to avoid calling the Principle an assumption at all: when compared with any specific, concrete assumption such as is involved in the radio-carbon method of dating, it hardly seems to qualify for the name.

5.4 *Uniformity as a principle of method*

Perhaps we can look at the Principle in a different light. There is a weaker claim, according to which we can speak of the very success of the sciences as showing the Uniformity of Nature. On this account of the matter, any general statement about the uniformity must remain very vague; but it will at any rate not pretend to be a presupposition, or assumption, or indeed anything which scientists could be described as making 'blindly'. In this sense of the phrase, there will be no room to deny that there is *some* uniformity in Nature: the fact that physical theories have been developed which have some application to the world, is all the evidence of this uniformity we need require—the Uniformity of Nature has been discovered, once for all. Even if the particular theories now accepted prove to have their weaknesses, that will not wipe out the successes already achieved: the existence of some degree of uniformity, in this sense, will be a fact beyond dispute. Thus interpreted, the Uniformity Principle is very unexciting, and we may prefer to pitch our standards of uniformity higher as we go along; but then, as the limitations of our present theories are discovered, that will be reason not so much for abandoning the physical sciences in despair as for developing more and better theories. At times, indeed, the Uniformity Principle has been treated almost as a manifesto, or as the statement of a programme: as if one said, "There are always uniformities which remain to be discovered." So understood, to say that physicists believe in the Uniformity of Nature will be to say, not that they have had some success in the past, nor that their present procedures are methodical; but rather that they are optimistic, and have hopes of getting somewhere in the future.

But, in whatever sense we understand the Uniformity Principle, whether as assumption, as discovery or as manifesto, it has one special weakness: that of irremediable vagueness. A principle stated in such general terms can be of no practical significance. For to talk of Nature as uniform without saying in what respect or to what degree it is uniform, is to say hardly anything: no one either assumes, or has discovered, or expects to discover an unlimited degree of uniformity in an unlimited number of respects.

The astrophysicist studying a new double-star, for instance, presumes not a general Uniformity but a particular and explicit similarity—namely, that this one double-star now under observation is comparable as a dynamical system with the sun and planets, in just such respects and to just such a degree as will entitle him to calculate the motions of its members by using Newton's inverse-square law. He does not need to presume anything more general than this: he does not assume, e.g., that *all* double-stars will turn out to be strictly comparable with the solar system—whether or no they are remains to be seen. A chronologist using the radio-carbon technique likewise assumes, not a general Uniformity, but a highly specific constancy in the atmospheric conditions since the time at which his specimen was formed. Whatever scientific problem one considers, one finds fresh assumptions and presumptions, all highly specific, and differing from one case to another. Nor indeed is it necessarily uniformities and correlations which are specially interesting. Non-uniformities and non-correlations, independencies and disconnections are quite as important, for instance, in discrediting old wives' tales and quack remedies.

As a result, it is impossible to state in any but completely formal terms a Principle of Uniformity common to all the sciences alike: different scientists working in different fields start off with different initial presumptions, and nothing more general will be of any use to them. There is nothing to prevent one's *saying*, "Scientists presume that, or have discovered that, or believe that Nature is Uniform", leaving what it is exactly that they presume, have discovered or believe entirely vague; but to say this is to make the very weakest of claims, which does no more than to indicate the form taken by scientists' presumptions, discoveries and ambitions. So if, in practice, scientists never seem to worry about the trustworthiness of their Uniformity Principle, that need be regarded neither as surprising nor as a sign of blindness on their part.

To conclude, then—the Principle of the Uniformity of Nature will not do the job designed for it by philosophers from Mill to Russell: being at best purely formal, it can serve as a premise in no physical arguments. But need it be any the less important for that? Might one not hold that the principle must be treated *as a principle*; that the inadequacies we have discovered come from forgetting this, and mistakenly treating it as a premise instead? Recognized for what it is, cannot a place in fact be found for it like that which we have seen is allotted to principles within physics?

The Rectilinear Propagation Principle, as we saw, has a place in physics for so long as the methods and arguments of geometrical optics are found of use: the abandonment of this principle would mean the end of geometrical optics as we know it. In a similar way, one can perhaps speak of all science as resting on certain formal principles, provided that these are recognized as being principles of policy, of method, or of 'reason', and not premises: abandoning these principles means the end, not of a single subject, but of science as we know it. Scientists certainly do, on occasion, invoke principles of this kind. If one reads the recent disputes over genetics, one finds Lysenko criticized not merely for failing to explain the observed facts and going against established theories, but even more for proceeding in an unscientific, unmethodical, if not a positively irrational manner. What repels scientists educated in the European tradition is the way in which he resorts to invective and ideological dogma to bolster his case: he seems to them to be attacking not just the particular theories they accept, but the very practice of rational scientific investigation.

If we interpret the idea of 'the uniformity of nature' in this particular way, the only question is, whether we should not replace it entirely by the idea of the uniformity of scientific procedures. Perhaps we ought. But it is worth recalling how "through all their logical apparatus" the principles of physics do "still speak about the world". The same may hold here: it is, after all, as a result of *experience* that we find out what are the rational ways of studying the world and its contents.

5.5 *Determinism: stuffs and substances*

We have remarked in several places on the differences between our everyday classification of stuffs, as 'wood', 'stone', 'water', etc., and the chemical classification of substances, as 'cadmium', 'sodium hydroxide' and the like. This difference becomes important if we consider one particular kind of experiment in which the claims of the Uniformity Principle are especially appealing. Suppose, for instance, that we take a cadmium vapour discharge lamp and pass an electric current through it: it will then emit its characteristic red spectrum line. So confident are we that this will happen that we feel that the lamp is, as it were, obliged to glow just so, and that its conforming to expectation is evidence of a uniformity in the properties of chemical substances. Furthermore, though it seems quite meaningful to suggest that, when we switched on again, a cadmium lamp might perfectly well glow differently, we believe this to be in the last degree unlikely, and our confidence that it will not do so seems again to be evidence that we are making some genuine assumption about chemical uniformity.

This example also brings to the fore another vexed question, that of determinism. It is easy to suppose that, when we talk of systems obeying the laws of physics and chemistry, the metaphor of obedience can be taken seriously, i.e. that the systems are in some way compelled by these laws to behave as they do. For instance, we may have the idea that, when the cadmium lamp is switched on, it cannot help but emit just such kinds of light as it does; and the same idea is attractive in the case of other phenomena—it seems, e.g., that the planets are constrained by the laws of dynamics, as by tramlines, to follow the elliptical paths they do. On this view, the more science advances, the more the Universe must be thought of as resembling a vast machine. Colour is given to this type of determinism by the physicist's use of the word 'must', and by the characteristic logic of nature-statements. For, using the methods of quantum mechanics, it is possible to infer from the atomic specification of cadmium that cadmium vapour *must* emit radiation of just such-and-such wavelengths when a current passes through it. So it seems that the lamp in our apparatus has no choice, poor thing: it *must* glow just as it does.

Two points can be made, which will help us to see how these conclusions are to be avoided. To begin with, we must not overlook all that goes on in the geological survey, and in the process of getting and refining, which take place before the lamp is ever constructed. As we saw before, if we forget what an astrophysicist *presumes* when he begins to study a double-star, we may be led to suppose, mistakenly, that he is making *assumptions* of a general kind about the constancy of laws of nature: so here, if we forget what geologists *presume* about the particular lumps of rock they unearth, we may be led to suppose, equally mistakenly, that chemists have to make general *assumptions* about the uniformity of the properties of chemical substances. For the question can always be raised, whether the manufacturers who delivered the metal from which our lamp was constructed were not mistaken in supplying it as pure cadmium; and this question is on a par with the question, whether the astrophysicist was right in presuming that the motion of the parts of his double-star was a purely gravitational phenomenon. It is no accident that we apply the adjective 'pure' both to kinds of chemical substance and to types of physical phenomenon. In their turn, the manufacturers rely on the surveying geologist and on their own testing procedures: they presume that ore from a correctly identified vein will, after a given process of refining and testing, yield an end-product which they will be entitled to sell as pure cadmium. If, however, the surveyor has made a mistake, or the vein was impure, or their tests gave deceptive results, it is always possible that the stuff they send out will contain other substances than cadmium, and even, though improbably, no cadmium at all.

The geologist responsible for identifying the vein of cadmium ore has, of course, his techniques for deciding when a stratum is of the composition required. Having identified a vein from tests on a sample, he will then expect the mass of rock from which the sample was taken to go on yielding the same substances, for as long as there is no reason to suspect changes in the composition of the rock. Our use of the word 'vein' helps to conceal this point. It is often left indeterminate whether a vein is to be identified by its texture, colour, etc. or by its chemical composition; and that is natural, for the first are taken as reliable signs of the second. Once again, nothing of a general nature can be said about what should lead him to suspect such changes: this is something which will depend entirely on the circumstances of any particular case. Furthermore, the geologist's presumptions will again be only initial ones. If anything goes wrong, he will reconsider them; and the first sign that something has gone wrong may be that a lamp, on being switched on, shines in a quite unexpected way. As with any initial presumption, the geologist will certainly be surprised if this happens, but he will not be desperate. He will take it, not as evidence of the breakdown of a Uniformity Principle, but rather as evidence of an undetected variation in the ore, and so of an unforeseen failure on the part of his surveying procedure: his presumptions, like the astrophysicist's presumptions, are both highly specific and open to rebuttal.

Our confidence that the lamp will glow in the way we have been led to expect accordingly reflects, not a general assumption about the uniformity of chemical substances, but rather a specific confidence that the chemists who supplied the material for the lamp sent what we ordered, and this in its turn depends on the geologist's highly specific presumption, that the next foot of ore will have effectively the same composition as the last hundred feet. It points also towards an important difference between everyday stuff-words and chemical substance-words, a difference worth comparing with that between generalizations and laws. For while a statement like "Wooden objects float" resembles in its logic the habit-statements of natural history, the statements of chemical theory, such as "Two molecules of hydrogen combine with one of oxygen to form two molecules of water", ($2H_2 + O_2 \rightarrow 2H_2O$), are as much nature-statements as the law of gravitation.

The consequences of this fact are crucial. First, and above all, it means that a chemical expression such as '$2H_2 + O_2 \rightarrow 2H_2O$' will be connected with the experimental results indirectly: like a law of nature, it will tell us about the world only if read in conjunction with other statements—in this case, directions for identifying such-a-stuff as qualifying for the chemical symbol 'H_2' and other stuffs as qualifying for the symbols 'O_2' and 'H_2O'. Accordingly, no experimental statements can be *deduced from* the chemical formula: rather, if we are given the chemical specification of the system under investigation, we can infer experimental conclusions by arguing *in accordance with* the formula.

This point is easily overlooked, since words which figure both within chemistry and outside it, such as 'water', 'iron' and 'salt' act as logical bridges: they are used sometimes as everyday stuff-words, sometimes as chemical substance-words, and often enough in a way which has about it something of both uses. Distinctions are not made in the sciences until they need to be; so, where the origin of a stuff carries with it a presumption as to its chemical nature, as is the case with water, one word may well be used to mark both origin and presumed nature. From the logician's point of view, however, there is one drawback: this double function helps to conceal the transition from the statements of theory to those of the laboratory, so that one may not notice that a distinction can be made between them.

Only from occasional remarks does the need to distinguish between the two uses of such words become clear, but these remarks are significant. The section on 'Water' in one well-known text-book of Inorganic Chemistry opens with the words, "Water is found in large quantities in the sea, rivers, etc." To the non-scientist this sentence is incurably comic: to the chemist, it is deadly serious. For the non-scientist reads it as he would the sentences, "Trout are found in large numbers in the streams of Dartmoor" and "There's gold in them thar hills." Read this way, it looks like a joke, since 'water' is what we *call* the stuff of the rivers and sea; so that to say this is as unhelpful as to say "I'll tell you what I've got in my pocket . . . its contents." In each case, what starts promisingly ends in bathos. The chemist, on the other hand, thinks of water more as a chemical substance than as an everyday stuff, and accordingly the trite-looking everyday sentence is transformed, for him, into the significant chemical statement, "Much of the stuff of which the sea and rivers consist can be counted as 'H_2O'." This sentence is far from tautologous: it is, indeed, a very necessary and practical piece of information, for only with this assurance can we confidently apply to the liquid we get from the sea the statements about 'H_2O' in books of chemical theory—such as that 'H_2O can be decomposed by electrolysis into H_2 and O_2'.

With this in mind, we can reconsider the case of the cadmium lamp. The source of our problem was this: one can infer from atomic theory that cadmium vapour *must* emit light of such-a-wavelength when a current passes through it, and it seemed that our cadmium lamp could hardly do anything else but conform to the general rule. Apparently, then, a chemist was justified in asserting, quite baldly, that our particular specimen *must* radiate just such wavelengths—that it was chemically compelled so to do. But now we can see part of the way out: the argument has both a major premise, 'Cadmium vapour when excited must emit such-and-such radiation', and a minor premise, 'This lamp contains cadmium vapour', and this minor premise is evidently not as trifling as it at first appeared. For, in so far as the stuff of which the lamp is made was identified by geological tests alone, there is no *necessity* that it shall satisfy also the chemical criteria, but only a presumption. This is the distinction which is hidden by our word 'vein'. The minor premise is, therefore, to be understood as saying, 'The stuff from which this lamp was made *can be counted as* cadmium', and the conclusion of the argument will accordingly be, not the bald 'This specimen *must* emit light of such-a-wavelength', but either '*In so far as* the stuff from which this lamp was made has been correctly identified as cadmium, it *must* emit light of such-a-wavelength', or else 'This specimen *will* emit such light.' Only in so far as the stuff can properly be counted as cadmium—or rather, to use the chemical symbol, as Cd—must it emit just that light. The minor premise is more than a simple class-membership statement: it is the essential identification-statement, without which there can be no bridge between a theoretical doctrine in chemistry and any experimental conclusion.

5.6 *Determinism: theoretical necessities are not constraints*

There is a second point to be considered, which also helps to remove the force of the deterministic doctrine, and connects with the things we noticed in an earlier chapter about the physicist's use of the word 'must'. When we use the word 'must', it is not always the thing which we say 'must do or be so-and-so' which is subject to compulsion, obligation or constraint. This is especially the case when the word 'must' is used in connection with inferences, and with the application of rules. You may, for instance, be struck by something in a girl's features and say, "Why, she must be Jack's sister"; but to say this is not to say that the girl is subject to any obligation or constraint to be Jack's sister. Rather, it is to show that something forces *you* to the conclusion that she is Jack's sister: it is you, not she, who is 'constrained'. Again, you may read in the newspaper that "32.000 people visited the Zoo on Sunday" and say, "They must mean 32,000 people, not 32": in this case also it is you, not they, who are 'driven'—driven, that is, to the conclusion that '32,000' was meant. The conclusions of arguments are very commonly expressed in such terms; and wherever this is done, the word 'must' marks the fact that the inferred conclusion has been drawn in a manner which could be justified by appeal to a rule of inference, law of nature, or generally accepted principle. Where there is such a rule of inference and a suitable set of premises, there can be only one conclusion: this, we say, 'must' be the conclusion—i.e. must be the proper conclusion to draw.

Inferences in the physical sciences, chemistry included, are no exception. When, for instance, we read off from a ray-diagram the depth which a shadow may in given circumstances be expected to have, we put our conclusion in the words, "So the shadow *must* be 10 ft. 6 in. deep." In saying this we are saying, not that there is any compulsion on the shadow to have just this depth—if indeed it means anything to speak of a shadow being under compulsion—but rather that, when one applies the methods of inference-drawing found reliable in such cases, there can be only one conclusion as to the depth the shadow can be expected to have. It is, accordingly, not the systems which physicists study which are forced by the Laws of Nature to be or do this or that, so much as physicists themselves: by accepting particular laws of nature as applicable in particular types of situation, they are required to draw just those conclusions about physical phenomena to which the laws lead, rather than others.

Perhaps the manner in which mathematical physics origin-

ally developed was, from a philosopher's point of view, unfortunate. The solar system, which provided the testing-ground
for the first coherent theories of dynamics, was too good an
example: the parallel between theory and fact was there so
close that the *long-term* prediction of actual developments came
to seem a more reasonable and practical aim than we are
entitled to expect. If there had been no such isolated, better-
than-laboratory prototype to study, we might have been less
inclined to overlook the steps involved in applying physical
theories. For, when one checks the motions of the planets
against the astronomer's dynamical calculations, it almost seems
that Newton's Laws are plain statements of fact about the
planets themselves: for a moment the logical gulf between
Kepler's Laws and Newton's seems to vanish. The tramlines of
our dynamical calculations are projected into the sky, and the
planets are seen to be running along them. In this mood, we
tend to think of the logical articulation of our mathematical
theories as having a physical counterpart in the Celestial Tramway along which the planets are constrained to move—the
Celestial Tramway being, so to speak, only the Inner Circle of
the Causal Nexus: determinism then seems an inescapable consequence of the success of our theory of planetary dynamics.
Alternatively, since there is manifestly no Tramway there in
fact, it is as though the 'laws of nature' had as counterparts
Divine Regulations, which the planets, being obedient creatures, conscientiously observed.

The Causal Nexus is, nevertheless, a myth. The necessities
of dynamics, and of all theories in the exact sciences, are of
another kind. It is not that the physicist believes the world to be
a machine, and that this premise is essential to the success of his
theories. Rather, the physicist develops as the central parts of
his theories techniques of inference-drawing and ways of
representing physical systems which can be used *inter alia* to
make exact predictions; and, if his inferences are to be regarded
as correct, they must be drawn by the use of the appropriate
techniques. Since only those theories are accepted which can
be made to fit accurately a considerable range of observed
phenomena, the results of correctly performed calculations can

thereafter be expected to fit the behaviour of appropriately chosen systems. But the 'must' which appears in any physical argument remains the 'must' of a correctly drawn inference, and we can read it into the conclusions we draw about actual collections of bodies or pieces of apparatus only by overlooking the fact that these conclusions depend on two presumptions: the overriding presumption that the theories employed are not in need of correction, and the particular presumption, that the system or apparatus being studied has been correctly identified ('placed') as falling within the scope of these theories.

The physicist's use of the word 'must' provides no warrant, therefore, for the idea that physics has proved that the Universe is a machine. On closer examination, in fact, one feature of this idea appears decidedly peculiar. For the machine of the determinist's picture is no ordinary machine—rather, it is the machine of an engineer's dreams. Likewise with all the paraphernalia of determinist metaphysics: causal chains, billiard balls and the rest.

All actual machines wear out. Their behaviour can be foretold from the design specification only for a limited time after manufacture; and further, it departs from the specified behaviour progressively, and in a more-or-less unpredictable manner, up to the moment of breakdown. What happens after that moment is quite unpredictable from the specification: it depends entirely on what the engineer decides to do with the broken-down machine. The determinist's machine, however, is unlike actual machines in this most characteristic respect: it did not occur to the nineteenth-century mechanist that the world-machine was liable to wear and tear. No wonder, for the machine he had in mind is the ideal machine, which will by definition behave for all time in the way laid down in the design specification. In this it betrays its origin. The determinist's machine, churning on to eternity with mathematical precision, bears the marks of its maker—it is the ghostly counterpart of our own mathematics: its mathematical precision reflects the rule-guided exactitude of the steps in our calculations.

Wittgenstein illustrated this point by considering the diagram, shown opposite, of a piston moving in a cylinder.

We are inclined to say that, in the machine here represented, as A rotates, B *must* move first one way, then the other. The piston, like the planets, seems compelled. But notice one thing: to say only this is to say nothing about any actual machine in the world. The diagram could be part of the specification of a number of possible machines, but it is not itself a machine: the 'must' of our conclusion can be read into statements about

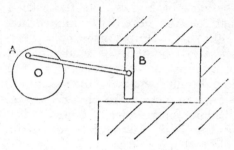

actual machines only by overlooking the vital minor premise, that the present state of some particular machine can be accurately represented by this diagram. All that we are entitled to infer from the diagram is this: that the more nearly the state of an actual machine can be so represented, the more closely can it be expected to move in the manner stated. Actual machines, being subject to wear and tear, do not conform to specification for an indefinitely long time: no single diagram such as this can, therefore, be accepted for ever as a reliable guide to their performance.

As with the determinist's world-machine, so with 'causal chains'. The chain manufacturer who made chains having the properties of causal chains would soon be a millionaire. For, as machines wear out and break down, so also do chains wear and snap. We do not find unbreakable chains in Nature, nor do we know how to make them. The unbreakable chain is an ideal, towards which our manufacturers work: actual chains are all liable to break and wear out, but the better the manufacturer succeeds in making them, the greater will be the strain which they will stand without breaking and the longer they will last. If causal chains seem so particularly tough and long-lasting, that is again a mark of their origin: the unbreakable causal chains in the determinist's picture of the world are unbreakable because they are the shadows cast by the logical chains of inferences in scientific arguments. If we do not find any unbreakable chains in Nature, this is a reminder that only exceptionally well-made artefacts behave according to specification for more than a limited time; and likewise it is only exceptional systems of bodies, like the solar system, whose behaviour continues for more than a limited time to be explicable in terms of a single, simple theory.

There is one last point to be made about causal chains, which connects with what was said earlier about the notion of a cause. The causal chains which are the metaphysical shadows of the arguments we employ in the physical sciences—the chains by which a shadow, for instance, is bound to be the depth it is —are not to be confused with the chains of circumstances which are features of the diagnostic sciences. They are distinct in two respects. First, the chains of circumstances which we speak of as leading, e.g., to a railway accident have nothing in the way of necessity about them: when we say "The chain of circumstances was as follows . . .", our aim is to tell how it was that the accident came to happen, not to show that it must inevitably have happened. It is a further question whether or no, under the circumstances, the accident was in any sense bound to happen. And secondly, such chains as these do not correspond one for one with causal chains. They could be fitted into the determinist's picture only by including in them links from many different causal chains—as many, indeed, as the different branches of scientific theory which would have to be invoked, if we were to produce an exhaustive picture of the physical processes involved in the accident. The ways in which things happen outside the laboratory conform with unlimited exactness neither to the decrees of a machine-like Destiny nor to the pattern of any one simple argument.

5.7 'Believing that . . .' and 'Regarding as . . .'

To leave the question of determinism at this point would, nevertheless, be unsatisfactory: one other thing urgently needs saying. Even supposing we grant that it is a mistake, for philosophical purposes, to project the logical necessities of our calculations into the sky, and to think that the planets are in any everyday sense constrained by them; nevertheless, it is a physicist's business to do something very like this. To the physicist, understanding why the planets move as they do means not simply having a mathematical theory with the help of which their orbits can be computed, but also being able to think of them in a way which makes sense of that theory. The scientist must be able, that is, to look at the systems of bodies he studies with a professional eye, and 'see' their behaviour in the way his theories require: this, as we have seen all along, is the purpose of using models in the physical sciences.

But though the use of models may at first look very like the determinist's philosophical mistake, the two things are in fact very different. To think that A *is* B is one thing, to think of A *as* B is another; and the Man Friday fallacy is one consequence of overlooking the difference between them. In some ways, indeed, the two things seem to be mutually exclusive, There is no room, e.g., to think of a cylinder of gas *as* a box full of fast-moving billiard balls, unless one knows very well that it is not in fact such a box. One cannot use the model of a box full of fast-moving billiard balls to explain the behaviour of a box full of fast-moving billiard balls: a model can only be used to explain the behaviour of things which are in fact distinct from it.[1] Nor is the physicist, whose explanation of the behaviour of gases requires us to think of a cylinder of gas as a box full of balls, in any danger of mistaking the one for the other—he knows very well the difference between them. Coming to understand the kinetic theory of gases does not involve coming to believe, in any everyday sense, that a cylinder of gas *is* such a box; and yet we do need, in learning the theory, to be able to look at the one as though it were the other, for only so shall we be able to use

---

[1] This is a logical point which crops up in connexion with many phrases containing the word 'as'. One can paint a picture of Mrs. Siddons *as* Ariadne, but a portrait of Mrs. Siddons is not 'a picture of Mrs. Siddons as Mrs. Siddons'.

the theory to understand the observed behaviour of hydrogen, oxygen, carbon dioxide and the rest.

What goes for the particular models of physics applies also to the physicist's general method of approach, of regarding his objects of study as articulated structures. In the physical sense of the phrase, one can hardly 'regard' the bones of the hand or the parts of an Anglepoise lamp 'as' articulated, for they *are* just that; and to explain the action of each it is sufficient to describe its method of articulation. But few natural objects are of this kind, and these few are mainly the concern of biologists; so that it is in accounting for the behaviour of systems which are in fact not articulated that the physicist has to look for as-it-were connexions, as-it-were structure, and as-it-were mechanism. Systems which are not in fact articulated structures are just those that he has to *regard as* articulated structures.

This point is sometimes hinted at by a distinction between the methodological determinist and the metaphysical determinist. It is suggested that the physicist does not need to *assert* that the world is a machine, i.e. that the behaviour of any system he chooses to study will prove to be as mechanical as the movements of, say, a steam-engine; but that he needs only to *assume*, for professional purposes, that everything in the particular field he is studying is determined and mechanical. This latter, tentative assumption is spoken of as methodological determinism, and contrasted with the more general and dogmatic, metaphysical determinism.

To put the point in this way is, however, still likely to mislead, for it conceals one essential feature of theoretical models. Remember: the physicist who uses the idea of light as a substance travelling does not *assume* for purposes of geometrical optics that light is literally travelling. To state his method in this way is to commit the Man Friday fallacy, for one can be said to *assume* only those things which could have been stated beforehand; whereas the point of this view of optical phenomena is, *inter alia*, that it brings with it a new way of talking and thinking about them, and only in this new way of talking can the so-called assumption even be stated. What the physicist does is, rather, to think of the old phenomena in this new way: shadows and the rest are now, for him, the consequences of something as-it-were travelling from the lamp to the illuminated object, though by all everyday criteria nothing need be travelling at all.

Nor does an astronomer *assume* that the planets are, literally, constrained by the inverse-square law to follow elliptical orbits: the Celestial Tramway is rather as it-were a tramway, the Connexions, Structure, Mechanism and Articulation of the Universe as-it-were connexions, structure, mechanism and articulation. If one speaks of the physicist's idea of mechanism as a provisional and professional assumption, people will be entitled to suppose that the advance of science may eventually prove the assumption justified. But there is no question of this: scientists will never be entitled to say to the public, "At last we are in a position, not merely to assume, but to announce definitely that the universe *is* a machine," any more than they will ever be able to say, "At last we have proved definitely that a hydrogen cylinder *is* a box full of fast-moving billiard balls." Models remain models, however far-reaching and fruitful their applications may become.

5.8 *Why popular physics misleads the layman*

One last point: the models of the theoretical sciences have parts to play not only on paper, but also in scientists' minds. And here one thing must be noticed about the way in which this book has been written: throughout, a great deal has deliberately been made explicit which in practice might often go unstated. Frequently one can see how a shadow comes to be the depth it is, without going to the length of drawing a ray-diagram: given the model of light as something which travels in straight lines, one can understand the phenomenon well enough and, having learnt to think of light in this way, one will often be able to dispense with all formulae and diagrams. Yet this fact does not mean that the formulae and diagrams are, logically speaking, any less central: if one were required to set out one's argument in full, or to explain the subject to a novice, it would be essential to use them. For the logician, therefore, such things as these, which in practice may sometimes be left unmentioned, are as important as the things which are always stated or worked out on paper; for our purposes, it has been necessary in every case to bring them into the open.

With this point in mind, we can return to a difficulty which we encountered at the very beginning of the book, and see the reason for it. There we noticed how easily misunderstandings arise when professional physicists set out to explain their theories to outsiders. The physicist says, "Heat is a form of motion", or "The universe is the three-dimensional surface of a four-dimensional balloon", or "A gas is a collection of minute particles moving with high velocities in all directions"; and in each case the onlooker either does not know what to understand by the pronouncement, or overlooks the unspoken, qualifying 'as it were' and so draws the wrong conclusions.

This sort of cross-purposes will perhaps be less surprising in the light of our subsequent discussion. For the physicist learns, as part of his training, to think and speak *in terms of* his theoretical models, and when he is required to popularize his subject he naturally turns to these for help. But to the outsider these theoretical models, however vivid, are neither familiar nor immediately intelligible, and their role is itself something which he needs to have explained. Inside physics, speaking within a theory and in terms of it, the scientist can do without the qualifying phrase 'as it were': he will perhaps see the implications of the kinetic theory for the gases he is studying in the laboratory the more clearly, the more vividly he can visualize gases as composed of minute billiard balls. In the laboratory, therefore, there will be every reason to say, "A gas is composed of . . ." instead of "A gas is, as it were, composed of. . . ." But when the scientist turns to speak to the outsider, the qualifying phrase becomes vitally necessary. After all, the gas is not in fact composed of minute billiard balls: the thing he has to explain is how physics is advanced by using billiard balls as a model in terms of which to think about gases.

There need be no mystery about this contrast. Often enough, a remark which is immediately intelligible in one situation will be either misleading or unintelligible in another. Thus, in the theatre, a member of the audience can whisper to his neighbour, "Here comes Cleopatra," as an actress comes on to the stage; and, provided the neighbour understands what is going on, he is in no danger of being misled. But the same words, used when passing the actress in the street next day, would be open to serious objection: it is in fact Edith Evans, not Cleopatra, we have met. Whether we may safely speak of 'Cleopatra', or must say rather 'as-it-were Cleopatra' ('the actress playing Cleopatra') depends entirely on the situation in which we are placed.

To explain the theories of physics in a manner which would be both genuinely intelligible to the outsider and free from risk of misunderstanding, a scientist must therefore reverse completely the language-shift to which he becomes accustomed in the course of his training, and use all the terms affected by the shift (such as 'force', 'energy', 'surface', 'billiard-ball', 'light', 'travel', 'structure', 'mechanism') in their everyday senses once again. Anything less than this will leave room for cross-purposes and misunderstandings of the old, deplorable kind.

One result of this reversal will be to increase the length of any account—though this is a small price to pay for understanding. We saw in an earlier chapter, for instance, how much longer a statement of Snell's Law must be if the technical vocabulary of 'light-rays' is eschewed, and the whole thing put in explicit terms. Where a physicist, among his colleagues, would describe the investigation leading up to the discovery of the law as 'an investigation of the optical properties of refracting media', the onlooker needs to think of it as 'seeing if a way can be found of extending the techniques of geometrical optics (ray-tracing, etc.) so as to be applicable when such things as shadows are formed under water, or the far side of a sheet of glass'. And whereas a physicist would state Snell's Law in the form, "The angle which the incident ray makes with the normal to the surface of the refracting medium (i) is related to the angle between the refracted ray and the normal (r) by the equation $\dfrac{\sin i}{\sin r} = \mu$", we laymen have to precede this statement by the preamble, "The techniques can be extended by altering the directions of the straight lines in our ray-diagram where they cross the surface, and thinking of light-rays as bending where they pass from one transparent medium to another, in such a way that . . ."; while the physicist's brief "The refractive index of water is 1.33" becomes "The constant in the equation governing the amount by which the lines in our ray-diagram are to be deflected when passing from one transparent medium to another is 1.33, for the transition from air to water."

This increase in length should have been foreseen. Physical scientists do not adopt their models and terminology for nothing, and greater conciseness of expression (what Mach calls 'economy') is one of the important advantages they aim at. But it imposes on the popularizer a duty which he is often tempted to ignore—to remember that theories draw their life from the phenomena they are used to explain, and to make sure that, in squeezing his account into a nutshell, he does not sacrifice first what he should retain till the very last: an adequate account of the physical phenomena in question, and of the manner in which the models used in the theory help the physicist to make sense of them.

SUGGESTED READING

INTRODUCTORY LEVEL

Philosophy of Science
Campbell, Norman, *What is Science?* (1921).

Mathematics and its Applications
Sawyer, W. W., *Mathematician's Delight* (Penguin ed., 1943).

Inductive Logic
Black, Max, *Critical Thinking* (1946), part III.

CLASSICAL DISCUSSIONS OF THE PHILOSOPHY OF SCIENCE
Galilei, Galileo, *Dialogue concerning the Two Principal Systems of the World* (1632, tr. 1661).
Newton, Isaac, *Mathematical Principles of Natural Philosophy* (1687, modern tr. Cajori, 1934).
Locke, John, *An Essay on Human Understanding* (1690).
Hume, David, *A Treatise of Human Nature* (1739).
Kant, Immanuel, *Critique of Pure Reason* (1781, 1787, modern tr. Kemp Smith, 1929).
All of these contain sections dealing with problems in the philosophy of science, and have greatly influenced the course of all subsequent discussion.

MODERN CLASSICS IN THE PHILOSOPHY OF SCIENCE
Whewell, William, *The Philosophy of the Inductive Sciences* (1840).
Mill, J. S., *A System of Logic* (1843), esp. Bk. III.
Mach, Ernst, *The Science of Mechanics* (1883, tr. 1907); Mach's essays on 'Economy' and 'Comparison' in *Popular Scientific Lectures* (1895) may also be recommended.
Hertz, Heinrich, *The Principles of Mechanics* (1894, tr. 1899), Introduction.
Poincaré, Henri, *Science and Hypothesis* (1902, tr. 1905).
Bridgman, P. W., *The Logic of Modern Physics* (1927).

OTHER GOOD GENERAL DISCUSSIONS
Born, Max, *Experiment and Theory in Physics* (1943).
Clifford, W. K., *The Common-sense of the Exact Sciences* (1885).

Clifford's essay on 'The Aims and Instruments of Scientific Thought' reprinted in *The Ethics of Belief* (1947) is excellent.

Dingle, Herbert, *Through Science to Philosophy* (1937).

Eddington, A. S., *The Nature of the Physical World* (1928, repr. Everyman ed.).

Einstein, A. and Infeld, L., *The Evolution of Physics* (1938).

Frank, Philipp, *Between Science and Philosophy* (1941).

Pearson, Karl, *The Grammar of Science* (1892, repr. Everyman ed.).

Planck, Max, *A Scientific Autobiography* (1948, tr. 1950).

Stebbing, L. S., *Philosophy and the Physicists* (1937, repr. Penguin ed.).

MORE ADVANCED DISCUSSIONS AND WORKS OF IMPORTANCE FOR PARTICULAR TOPICS

Campbell, Norman, *Physics, the Elements* (1920).

Dingler, Hugo, *Die Methode der Physik* (1938).

Eddington, A. S., *The Philosophy of Physical Science* (1939).

(ed. Schilpp), *Albert Einstein, Philosopher-Scientist* (1949).

Kneale, William, *Probability and Induction* (1949).

Popper, K. R., *Logik der Forschung* (1935).

Ramsey, F. P., *The Foundations of Mathematics* (1931).

Ryle, Gilbert, *The Concept of Mind* (1949).

Schlick, Moritz, *Gesammelte Aufsätze* (1938).

Watson, W. H., *On Understanding Physics* (1938).

Whitrow, G. J., *The Structure of the Universe* (Hutchinson's University Library, 1949).

Wittgenstein, L., *Tractatus Logico-Philosophicus* (1922), esp. section 6.3 ff.

Woodger, J. H., *Biological Principles* (1929).

The *British Journal for the Philosophy of Science*, published quarterly, contains important articles on a variety of subjects: those in early issues by Prof. H. Dingle can be particularly recommended. The Penguin series *Science News* also contains worth-while articles on the philosophy of science from time to time. From more out-of-the-way periodicals, two papers are worth special mention, Prof. G. G. Simpson's article on classification in taxonomy (*Bulletin of the American Museum for Natural History*, 1945) and Prof. K. R. Popper's article on the part played by tradition in science (*Rationalist Annual*, 1949).

INDEX

ABSOLUTE Zero of Temperature, 129 ff
Accuracy, degrees of, 29, 70 ff, 111, 113
Action at a distance, 118
Aristotle, 46, 117, 118
Atomic model, 12, 39, 40, 137–9

BACON, Roger, 64
Bent stick phenomenon, 61–2, 149–50.
Bergson, H., 125
Black, Max, 171
Bohr, N., 138
Born, Max, 13, 123, 171
Boyle's law, 86–8, 111
Bridgman, P. W., 171
Brownian motion, 137–8

CALORIC fluid, 39, 137
Campbell, Norman, 171, 172
Carroll, Lewis, 102
Causal chains, 119, 124, 162, 163–4
    connexions, 41, 54
    nexus, 124, 161
Causality, 9, 119, 123
Cause, notion of, 10, 119 ff.
Charles' law, 131
Chemical substances, 53
    and stuffs, 51 n., 146
    uniformity of, 154 ff.
Chemistry, 46, 137
Churchill, Winston, 72
Classification, taxonomic, 50–2, 145–6

Cleopatra, 169
Clifford, W. K., 171–2
Cloud chamber, 136, 139
Complete description, 118–9, 124
Confirmation, 110
    theory of, 112–3
Constant conjunction, 91, 96, 98, 103
Contours, existence of, 135, 137
Conventionalism, 75, 83, 88
Cosmic epoch, 91, 99
Crusoe, Robinson, 19–20, 134

DALTON, J., 46
Deduction, role of, 41, 84, 106 ff.
Deductive systems, 62, 77 ff., 84
Description and explanation, 53 ff.
Determinism, 94, 154 ff.
    metaphysical and methodological, 166
Diagnostic sciences, 121 ff.
Diffraction, 29, 68, 69
Dingle, H., 172
Dingler, H., 172
Discovery and inference, 19, 24–5, 42, 43 n., 76–7
    accidental, 44
Dynamics, Newtonian, 33, 46, 70, 118
    Aristotelian, 46, 118

EDDINGTON, Sir A., 11–12, 103, 108, 124 ff., 172
Einstein, Albert, 13, 15, 16, 38, 43, 70, 85, 117–19, 123–4, 137, 144, 172

Empirical character of science, 80–1
Ether, 38, 118, 137
Exactitude, mathematical, 70 ff.
Exactness, practical, 70 ff.
Euclidean definition of point, 72
  straight lines, 71
Evans, Dame Edith, 169
Experiments, 49, 57, 65 ff., 73 ff., 110, 111

FRANK, P., 172

GALILEO, 130, 171
Geiger-counter, 136

HABIT statements, 50, 85, 157
Heisenberg, W., 124
Heraclitus, 22
Hertz, H., 171
Horizon of science, 117–19, 123–4
Hume, D., 91–6, 103, 171
Hypotheses, 11, 49, 79, 80–3

IDEAL Gas, notion of, 131
Ideals, theoretical, 70 ff.
Identification in chemistry, 104, 147, 155–9, 162
Imagination, 43
Induction, 9, 43, 140
Inference, syllogistic, 33, 49, 102
Inference tickets, 93–4, 103
Inferring techniques, 23 ff., 30, 33, 58, 61, 64, 93, 95, 128, 160, 161–2

JEANS, Sir J., 11, 12–13, 108

KANT, I., 128–9, 171
Kepler, 64
Kepler's Laws, 87–8, 161

Kinetic theory of matter, 39, 165–6, 168
Kneale, W., 91, 98, 101–2, 138–9, 172

LANGUAGE everyday, 18–19
  onlookers and participants, 13, 58, 169–70
  scientific, 13, 21
    conciseness of, 15, 170
    and everyday, 35–6, 47, 50 ff., 105, 145, 169
Language shift, 13, 169
Laplace, 117
Law-like statements, 78
Laws, scope of, 31, 63
  and principles, 83–4
  phenomenological, 86–8
Laws of Nature, 11, 49, 52, 57 ff.
  and generalizations, 10, 34, 77, 99, 105, 110, 126, 141–2, 144–6, 151
  as maxims, 100 ff.
  cf. laws of projection, 109–10
  logical character of, 11, 78 ff., 90 ff.
  not 'true', 78, 101
Leibniz, 38, 55, 118
Light, everyday view of, 21–3, 26
  Greek view of, 23, 26, 30, 39
  rectilinear propagation of, 17 ff., 23 ff., 29–30, 57, 71, 83, 86, 109, 138, 154
  wave theory of, 94, 113 ff.
Light-ray, idea of, 26–9, 60–1, 65, 69, 70–2, 77–8, 114, 126, 130–1
Locke, 91–2, 103, 171
Lysenko, 154

MACH, E., 40–3, 54, 84, 91–6, 105 ff., 150, 170, 171

Man Friday, 23, 135
    fallacy, 20, 40, 138–9, 165, 166
Maps and itineraries, 121–3
    and methods of projection, 127, 132–3
Mathematics, role of, 11, 26, 31–2, 70, 108, 128, 130, 162, 165 ff.
    world of, 33
Maxwell, J. C., 117
    principles of electromagnetism, 86
Mill, J. S., 141, 153, 171
    Methods, 9, 119
Models, 11, 12, 29, 30, 34–5, 39, 165–6, 167–9
    fertility of, 37, 50
Motion, equations of, 33, 109

NATURAL HISTORY, 34
    and physics, 44 ff., 50 ff., 55, 67, 74, 82–3, 85, 111–2, 141–2, 145
Nature-statements, 50, 87, 155, 157
Necessity and laws of nature, 91, 92, 96, 103
    in physics, 159 ff.
Newton, Isaac, 46, 117, 171
    Laws of Motion, 86, 88, 90, 161
    Law of Gravitation, 99–100, 145, 153

OBSERVATIONS, 54
Optical homogeneity, 63, 74, 77
Optics, geometrical, 17 ff., 23 ff., 36, 57 ff., 65–6, 69, 83, 108, 113 ff., 127, 154, 166
    physical, 36–8, 69, 89, 113 ff.
Ostwald, W., 138

PARTICLE, idea of, 72
Pearson, Karl, 172
Phenomenalism, 40–1, 105 ff.
Phlogiston, 81, 137
Physics, popularization of, 11 ff., 108, 167 ff.
Planck, M., 38, 172
Poincaré, H., 10, 171
Point, idea of, 72
Popper, K., 54, 172
Principles and laws, 83–4, 86
Probability calculus, 10, 49, 112–3
Procrustes, 126
Protons & electrons, mass-ratio and number of, 125
Ptolemy, 64

QUANTUM mechanics, 35, 113, 118–19, 123–4

RADIO-CARBON dating, 151, 153
Ramsey, F. P., 91–2, 100 ff., 172
Ray-diagrams, 25–6, 108, 127–8, 167
Refraction, 27, 29, 55, 58 ff., 63 ff, 73 ff., 169
    anomalous, 60, 64, 77, 79
Refractive index, 60, 63, 80, 85, 170
Regularities, form of, 44–5, 64, 77
Relativity, general theory of, 12, 85
Representation, methods of, 32, 115 ff., 122, 126 ff.,
    of phenomena, 27, 29, 41–2
Römer, 37
Russell, B., 124, 141, 153
Ryle, G., 78, 93–4, 103, 172

SAWYER, W. W., 171
Scattering, 71

Schlick, M., 91–2, 100 ff., 172
Scope, 31, 63, 69, 78
    of theories, 112–13
Simpson, G. G., 172
Snell's law, 58, 63 ff., 70, 73 ff., 77-8, 85, 86, 109, 114, 169
Stark effect, 75
Statistical mechanics, 113
Stebbing, L. S., 172
Stuff and chemical substance, 51 n., 146, 154 ff.
Syllogism, 33, 49, 102
System, 47, 77, 146

TEMPERATURE, 128, 129 ff.
    ideal-gas scale of, 131–3
    logarithmic scale of, 132–3
Thermodynamics, principles of, 86
Theories, basic, 113 ff., 123
    Pyramid model for, 84
    Stratification of, 80–1
Theoretical entities, 11
    existence of, 38, 134 ff.
Theoretical ideals, 70 ff.
Thomson, J. J., 138
Time and causation, 121

Truth and laws of Nature, 78–9, 86–7, 98, 99, 101, 112
    and theories, 114 ff., 128

UNIFORMITY of chemical substances, 154 ff.
    of Nature, 9, 140 ff.
Universe as a machine, 155, 162–4, 167
    spherical model of, 12, 15–16, 40, 168

WAISMANN, F. 117
Watson, W. H., 172
Whewell, W., 171
White, Gilbert, 54
Whitehead, A., 91, 98 ff.
Whitman, W., 22
Whitrow, G. J., 172
Wittgenstein, L., 13–14, 51, 81, 88–9, 129, 162–3, 172
Woodger, J. H., 172

ZEEMAN effect, 75

Revised January, 1970

harper ✦ torchbooks

American Studies: General

HENRY ADAMS Degradation of the Democratic Dogma. ‡ *Introduction by Charles Hirschfeld.* TB/1450

LOUIS D. BRANDEIS: Other People's Money, *and How the Bankers Use It. Ed. with Intro, by Richard M. Abrams* TB/3081

HENRY STEELE COMMAGER, Ed.: The Struggle for Racial Equality TB/1300

CARL N. DEGLER: Out of Our Past: *The Forces that Shaped Modern America* CN/2

CARL N. DEGLER, Ed.: Pivotal Interpretations of American History
Vol. I TB/1240; Vol. II TB/1241

A. S. EISENSTADT, Ed.: The Craft of American History: *Selected Essays*
Vol. I TB/1255; Vol. II TB/1256

LAWRENCE H. FUCHS, Ed.: American Ethnic Politics TB/1368

MARCUS LEE HANSEN: The Atlantic Migration: 1607-1860. *Edited by Arthur M. Schlesinger. Introduction by Oscar Handlin* TB/1052

MARCUS LEE HANSEN: The Immigrant in American History. *Edited with a Foreword by Arthur M. Schlesinger* TB/1120

ROBERT L. HEILBRONER: The Limits of American Capitalism TB/1305

JOHN HIGHAM, Ed.: The Reconstruction of American History TB/1068

ROBERT H. JACKSON: The Supreme Court in the American System of Government TB/1106

JOHN F. KENNEDY: A Nation of Immigrants. *Illus. Revised and Enlarged. Introduction by Robert F. Kennedy* TB/1118

LEONARD W. LEVY, Ed.: American Constitutional Law: *Historical Essays* TB/1285

LEONARD W. LEVY, Ed.: Judicial Review and the Supreme Court TB/1296

LEONARD W. LEVY: The Law of the Commonwealth and Chief Justice Shaw: *The Evolution of American Law, 1830-1860* TB/1309

GORDON K. LEWIS: Puerto Rico: *Freedom and Power in the Caribbean. Abridged edition* TB/1371

HENRY F. MAY: Protestant Churches and Industrial America TB/1334

RICHARD B. MORRIS: Fair Trial: *Fourteen Who Stood Accused, from Anne Hutchinson to Alger Hiss* TB/1335

GUNNAR MYRDAL: An American Dilemma: *The Negro Problem and Modern Democracy. Introduction by the Author.*
Vol. I TB/1443; Vol. II TB/1444

GILBERT OSOFSKY, Ed.: The Burden of Race: *A Documentary History of Negro-White Relations in America* TB/1405

CONYERS READ, Ed.: The Constitution Reconsidered. *Revised Edition. Preface by Richard B. Morris* TB/1384

ARNOLD ROSE: The Negro in America: *The Condensed Version of Gunnar Myrdal's* An American Dilemma. *Second Edition* TB/3048

JOHN E. SMITH: Themes in American Philosophy: *Purpose, Experience and Community* TB/1466

WILLIAM R. TAYLOR: Cavalier and Yankee: *The Old South and American National Character* TB/1474

American Studies: Colonial

BERNARD BAILYN: The New England Merchants in the Seventeenth Century TB/1149

ROBERT E. BROWN: Middle-Class Democracy and Revolution in Massachusetts, 1691-1780. *New Introduction by Author* TB/1413

JOSEPH CHARLES: The Origins of the American Party System TB/1049

HENRY STEELE COMMAGER & ELMO GIORDANETTI, Eds.: Was America a Mistake? *An Eighteenth Century Controversy* TB/1329

WESLEY FRANK CRAVEN: The Colonies in Transition: 1660-1712† TB/3084

CHARLES GIBSON: Spain in America † TB/3077

CHARLES GIBSON, Ed.: The Spanish Tradition in America + HR/1351

LAWRENCE HENRY GIPSON: The Coming of the Revolution: 1763-1775. † *Illus.* TB/3007

JACK P. GREENE, Ed.: Great Britain and the American Colonies: 1606-1763. + *Introduction by the Author* HR/1477

AUBREY C. LAND, Ed.: Bases of the Plantation Society + HR/1429

JOHN LANKFORD, Ed.: Captain John Smith's America: *Selections* from his Writings ‡ TB/3078

LEONARD W. LEVY: Freedom of Speech and Press in Early American History: *Legacy of Suppression* TB/1109

† The New American Nation Series, edited by Henry Steele Commager and Richard B. Morris.
‡ American Perspectives series, edited by Bernard Wishy and William E. Leuchtenburg.
α History of Europe series, edited by J. H. Plumb.
§ The Library of Religion and Culture, edited by Benjamin Nelson.
‖ Researches in the Social, Cultural, and Behavioral Sciences, edited by Benjamin Nelson.
Σ Harper Modern Science Series, edited by James A. Newman.
° Not for sale in Canada.
+ Documentary History of the United States series, edited by Richard B. Morris.
# Documentary History of Western Civilization series, edited by Eugene C. Black and Leonard W. Levy.
▲ The Economic History of the United States series, edited by Henry David et al.
¶ European Perspectives series, edited by Eugene C. Black.
** Contemporary Essays series, edited by Leonard W. Levy.
* The Stratum Series, edited by John Hale.

PERRY MILLER: Errand Into the Wilderness
TB/1139
PERRY MILLER & T. H. JOHNSON, Eds.: The Puritans: *A Sourcebook of Their Writings*
Vol. I TB/1093; Vol. II TB/1094
EDMUND S. MORGAN: The Puritan Family: *Religion and Domestic Relations in Seventeenth Century New England*
TB/1227
RICHARD B. MORRIS: Government and Labor in Early America
TB/1244
WALLACE NOTESTEIN: The English People on the Eve of Colonization: 1603-1630. † *Illus.*
TB/3006
FRANCIS PARKMAN: The Seven Years War: *A Narrative Taken from Montcalm and Wolfe, The Conspiracy of Pontiac, and A Half-Century of Conflict. Edited by John H. McCallum*
TB/3083
LOUIS B. WRIGHT: The Cultural Life of the American Colonies: 1607-1763. † *Illus.*
TB/3005
YVES F. ZOLTVANY, Ed.: The French Tradition in America +
HR/1425

American Studies: The Revolution to 1860

JOHN R. ALDEN: The American Revolution: 1775-1783. † *Illus.*
TB/3011
MAX BELOFF, Ed.: The Debate on the American Revolution, 1761-1783: *A Sourcebook*
TB/1225
RAY A. BILLINGTON: The Far Western Frontier: 1830-1860. † *Illus.*
TB/3012
STUART BRUCHEY: The Roots of American Economic Growth, 1607-1861: *An Essay in Social Causation. New Introduction by the Author.*
TB/1350
WHITNEY R. CROSS: The Burned-Over District: *The Social and Intellectual History of Enthusiastic Religion in Western New York, 1800-1850*
TB/1242
NOBLE E. CUNNINGHAM, JR., Ed.: The Early Republic, 1789-1828 +
HR/1394
GEORGE DANGERFIELD: The Awakening of American Nationalism, 1815-1828. † *Illus.*
TB/3061
CLEMENT EATON: The Freedom-of-Thought Struggle in the Old South. *Revised and Enlarged. Illus.*
TB/1150
CLEMENT EATON: The Growth of Southern Civilization, 1790-1860. † *Illus.*
TB/3040
ROBERT H. FERRELL, Ed.: Foundations of American Diplomacy, 1775-1872 +
HR/1393
LOUIS FILLER: The Crusade against Slavery: 1830-1860. † *Illus.*
TB/3029
DAVID H. FISCHER: The Revolution of American Conservatism: *The Federalist Party in the Era of Jeffersonian Democracy*
TB/1449
WILLIAM W. FREEHLING, Ed.: The Nullification Era: *A Documentary Record* ‡
TB/3079
WILLIM W. FREEHLING: Prelude to Civil War: *The Nullification Controversy in South Carolina, 1816-1836*
TB/1359
PAUL W. GATES: The Farmer's Age: *Agriculture, 1815-1860* ∆
TB/1398
FELIX GILBERT: The Beginnings of American Foreign Policy: *To the Farewell Address*
TB/1200
ALEXANDER HAMILTON: The Reports of Alexander Hamilton. ‡ *Edited by Jacob E. Cooke*
TB/3060
THOMAS JEFFERSON: Notes on the State of Virginia. ‡ *Edited by Thomas P. Abernethy*
TB/3052
FORREST MCDONALD, Ed.: Confederation and Constitution, 1781-1789 +
HR/1396

BERNARD MAYO: Myths and Men: *Patrick Henry, George Washington, Thomas Jefferson*
TB/1108
JOHN C. MILLER: Alexander Hamilton and the Growth of the New Nation
TB/3057
JOHN C. MILLER: The Federalist Era: 1789-1801. † *Illus.*
TB/3027
RICHARD B. MORRIS, Ed.: Alexander Hamilton and the Founding of the Nation. *New Introduction by the Editor*
TB/1448
RICHARD B. MORRIS: The American Revolution Reconsidered
TB/1363
CURTIS P. NETTELS: The Emergence of a National Economy, 1775-1815 ∆
TB/1438
DOUGLASS C. NORTH & ROBERT PAUL THOMAS, Eds.: *The Growth of the American Economy to 1860* +
HR/1352
R. B. NYE: The Cultural Life of the New Nation: 1776-1830. † *Illus.*
TB/3026
GILBERT OSOFSKY, Ed.: Puttin' On Ole Massa: *The Slave Narratives of Henry Bibb, William Wells Brown, and Solomon Northup* ‡
TB/1432
JAMES PARTON: The Presidency of Andrew Jackson. *From Volume III of the* Life *of* Andrew Jackson. *Ed. with Intro. by Robert V. Remini*
TB/3080
FRANCIS S. PHILBRICK: The Rise of the West, 1754-1830. † *Illus.*
TB/3067
MARSHALL SMELSER: The Democratic Republic, 1801-1815 †
TB/1406
TIMOTHY L. SMITH: Revivalism and Social Reform: *American Protestantism on the Eve of the Civil War*
TB/1229
JACK M. SOSIN, Ed.: The Opening of the West +
HR/1424
GEORGE ROGERS TAYLOR: The Transportation Revolution, 1815-1860 ∆
TB/1347
A. F. TYLER: Freedom's Ferment: *Phases of American Social History from the Revolution to the Outbreak of the Civil War. Illus.*
TB/1074
GLYNDON G. VAN DEUSEN: The Jacksonian Era: 1828-1848. † *Illus.*
TB/3028
LOUIS B. WRIGHT: Culture on the Moving Frontier
TB/1053

American Studies: The Civil War to 1900

W. R. BROCK: An American Crisis: *Congress and Reconstruction, 1865-67* °
TB/1283
T. C. COCHRAN & WILLIAM MILLER: The Age of Enterprise: *A Social History of Industrial America*
TB/1054
W. A. DUNNING: Reconstruction, Political and Economic: 1865-1877
TB/1073
HAROLD U. FAULKNER: Politics, Reform and Expansion: 1890-1900. † *Illus.*
TB/3020
GEORGE M. FREDRICKSON: The Inner Civil War: *Northern Intellectuals and the Crisis of the Union*
TB/1358
JOHN A. GARRATY: The New Commonwealth, 1877-1890 +
TB/1410
JOHN A. GARRATY, Ed.: The Transformation of American Society, 1870-1890 +
HR/1395
WILLIAM R. HUTCHISON, Ed.: American Protestant Thought: *The Liberal Era* ‡
TB/1385
HELEN HUNT JACKSON: A Century of Dishonor: *The Early Crusade for Indian Reform.* † *Edited by Andrew F. Rolle*
TB/3063
ALBERT D. KIRWAN: Revolt of the Rednecks: *Mississippi Politics, 1876-1925*
TB/1199
WILLIAM G. MCLOUGHLIN, Ed.: The American Evangelicals, 1800-1900: An Anthology ‡
TB/1382
ARTHUR MANN: Yankee Reforms in the Urban Age: *Social Reform in Boston, 1800-1900*
TB/1247

RNOLD M. PAUL: Conservative Crisis and the Rule of Law: *Attitudes of Bar and Bench, 1887-1895. New Introduction by Author*
TB/1415

AMES S. PIKE: The Prostrate State: *South Carolina under Negro Government.* ‡ *Intro. by Robert F. Durden*
TB/3085

WHITELAW REID: After the War: *A Tour of the Southern States, 1865-1866.* ‡ *Edited by C. Vann Woodward*
TB/3066

FRED A. SHANNON: The Farmer's Last Frontier: ...*Agriculture, 1860-1897*
TB/1348

VERNON LANE WHARTON: The Negro in Mississippi, 1865-1890
TB/1178

American Studies: The Twentieth Century

RICHARD M. ABRAMS, Ed.: The Issues of the Populist and Progressive Eras, 1892-1912 +
HR/1428

RAY STANNARD BAKER: Following the Color Line: *American Negro Citizenship in Progressive Era.* ‡ *Edited by Dewey W. Grantham, Jr. Illus.*
TB/3053

RANDOLPH S. BOURNE: War and the Intellectuals: *Collected Essays, 1915-1919.* ‡ *Edited by Carl Resek*
TB/3043

A. RUSSELL BUCHANAN: The United States and World War II. † *Illus.*
Vol. I TB/3044; Vol. II TB/3045

THOMAS C. COCHRAN: The American Business System: *A Historical Perspective, 1900-1955*
TB/1080

FOSTER RHEA DULLES: America's Rise to World Power: 1898-1954. † *Illus.*
TB/3021

JEAN-BAPTISTE DUROSELLE: From Wilson to Roosevelt: *Foreign Policy of the United States, 1913-1945. Trans. by Nancy Lyman Roelker*
TB/1370

HAROLD U. FAULKNER: The Decline of Laissez Faire, 1897-1917
TB/1397

JOHN D. HICKS: Republican Ascendancy: 1921-1933. † *Illus.*
TB/3041

ROBERT HUNTER: Poverty: *Social Conscience in the Progressive Era.* ‡ *Edited by Peter d'A. Jones*
TB/3065

WILLIAM E. LEUCHTENBURG: Franklin D. Roosevelt and the New Deal: 1932-1940. † *Illus.*
TB/3025

WILLIAM E. LEUCHTENBURG, Ed.: The New Deal: *A Documentary History* +
HR/1354

ARTHUR S. LINK: Woodrow Wilson and the Progressive Era: 1910-1917. † *Illus.* TB/3023

BROADUS MITCHELL: Depression Decade: *From New Era through New Deal, 1929-1941* ∆
TB/1439

GEORGE E. MOWRY: The Era of Theodore Roosevelt and the Birth of Modern America: 1900-1912. † *Illus.*
TB/3022

WILLIAM PRESTON, JR.: Aliens and Dissenters: *Federal Suppression of Radicals, 1903-1933*
TB/1287

WALTER RAUSCHENBUSCH: Christianity and the Social Crisis. ‡ *Edited by Robert D. Cross*
TB/3059

GEORGE SOULE: Prosperity Decade: *From War to Depression, 1917-1929* ∆
TB/1349

GEORGE B. TINDALL, Ed.: A Populist Reader: *Selections from the Works of American Populist Leaders*
TB/3069

TWELVE SOUTHERNERS: I'll Take My Stand: *The South and the Agrarian Tradition. Intro. by Louis D. Rubin, Jr.; Biographical Essays by Virginia Rock*
TB/1072

Art, Art History, Aesthetics

CREIGHTON GILBERT, Ed.: Renaissance Art ** *Illus.*
TB/1465

EMILE MALE: The Gothic Image: *Religious Art in France of the Thirteenth Century.* § 190 illus.
TB/344

MILLARD MEISS: Painting in Florence and Siena After the Black Death: *The Arts, Religion and Society in the Mid-Fourteenth Century.* 169 illus.
TB/1148

ERWIN PANOFSKY: Renaissance and Renascences in Western Art. *Illus.*
TB/1447

ERWIN PANOFSKY: Studies in Iconology: *Humanistic Themes in the Art of the Renaissance. 180 illus.*
TB/1077

JEAN SEZNEC: The Survival of the Pagan Gods: *The Mythological Tradition and Its Place in Renaissance Humanism and Art. 108 illus.*
TB/2004

OTTO VON SIMSON: The Gothic Cathedral: *Origins of Gothic Architecture and the Medieval Concept of Order. 58 illus.*
TB/2018

HEINRICH ZIMMER: Myths and Symbols in Indian Art and Civilization. *70 illus.* TB/2005

Asian Studies

WOLFGANG FRANKE: China and the West: *The Cultural Encounter, 13th to 20th Centuries. Trans. by R. A. Wilson*
TB/1326

L. CARRINGTON GOODRICH: A Short History of the Chinese People. *Illus.*
TB/3015

DAN N. JACOBS, Ed.: The New Communist Manifesto and Related Documents. *3rd revised edn.*
TB/1078

DAN N. JACOBS & HANS H. BAERWALD, Eds.: Chinese Communism: *Selected Documents*
TB/3031

BENJAMIN I. SCHWARTZ: Chinese Communism and the Rise of Mao
TB/1308

BENJAMIN I. SCHWARTZ: In Search of Wealth and Power: *Yen Fu and the West* TB/1422

Economics & Economic History

C. E. BLACK: The Dynamics of Modernization: *A Study in Comparative History* TB/1321

STUART BRUCHEY: The Roots of American Economic Growth, 1607-1861: *An Essay in Social Causation. New Introduction by the Author.*
TB/1350

GILBERT BURCK & EDITORS OF *Fortune*: The Computer Age: *And its Potential for Management*
TB/1179

JOHN ELLIOTT CAIRNES: The Slave Power. ‡ *Edited with Introduction by Harold D. Woodman*
TB/1433

SHEPARD B. CLOUGH, THOMAS MOODIE & CAROL MOODIE, Eds.: Economic History of Europe: *Twentieth Century* #
HR/1388

THOMAS C. COCHRAN: The American Business System: *A Historical Perspective, 1900-1955*
TB/1180

ROBERT A. DAHL & CHARLES E. LINDBLOM: Politics, Economics, and Welfare: *Planning and Politico-Economic Systems Resolved into Basic Social Processes*
TB/3037

PETER F. DRUCKER: The New Society: *The Anatomy of Industrial Order* TB/1082

HAROLD U. FAULKNER: The Decline of Laissez Faire, 1897-1917 ∆
TB/1397

PAUL W. GATES: The Farmer's Age: *Agriculture, 1815-1860* ∆
TB/1398

WILLIAM GREENLEAF, Ed.: American Economic Development Since 1860 +
HR/1353

J. L. & BARBARA HAMMOND: The Rise of Modern Industry. || *Introduction by R. M. Hartwell*
TB/1417

ROBERT L. HEILBRONER: The Future as History: *The Historic Currents of Our Time and the Direction in Which They Are Taking America* TB/1386
ROBERT L. HEILBRONER: The Great Ascent: *The Struggle for Economic Development in Our Time* TB/3030
FRANK H. KNIGHT: The Economic Organization TB/1214
DAVID S. LANDES: Bankers and Pashas: *International Finance and Economic Imperialism in Egypt. New Preface by the Author* TB/1412
ROBERT LATOUCHE: The Birth of Western Economy: *Economic Aspects of the Dark Ages* TB/1290
ABBA P. LERNER: Everbody's Business: *A Reexamination of Current Assumptions in Economics and Public Policy* TB/3051
W. ARTHUR LEWIS: Economic Survey, 1919-1939 TB/1446
W. ARTHUR LEWIS: The Principles of Economic Planning. *New Introduction by the Author°* TB/1436
ROBERT GREEN MC CLOSKEY: American Conservatism in the Age of Enterprise TB/1137
PAUL MANTOUX: The Industrial Revolution in the Eighteenth Century: *An Outline of the Beginnings of the Modern Factory System in England°* TB/1079
WILLIAM MILLER, Ed.: Men in Business: *Essays on the Historical Role of the Entrepreneur* TB/1081
GUNNAR MYRDAL: An International Economy. *New Introduction by the Author* TB/1445
HERBERT A. SIMON: The Shape of Automation: *For Men and Management* TB/1245
PERRIN STRYER: The Character of the Executive: *Eleven Studies in Managerial Qualities* TB/1041
RICHARD S. WECKSTEIN, Ed.: Expansion of World Trade and the Growth of National Economies ** TB/1373

Education

JACQUES BARZUN: The House of Intellect TB/1051
RICHARD M. JONES, Ed.: Contemporary Educational Psychology: *Selected Readings* ** TB/1292
CLARK KERR: The Uses of the University TB/1264

Historiography and History of Ideas

HERSCHEL BAKER: The Image of Man: *A Study of the Idea of Human Dignity in Classical Antiquity, the Middle Ages, and the Renaissance* TB/1047
J. BRONOWSKI & BRUCE MAZLISH: The Western Intellectual Tradition: *From Leonardo to Hegel* TB/3001
EDMUND BURKE: On Revolution. Ed. by Robert A. Smith TB/1401
WILHELM DILTHEY: Pattern and Meaning in History: *Thoughts on History and Society.° Edited with an Intro. by H. P. Rickman* TB/1075
ALEXANDER GRAY: The Socialist Tradition: *Moses to Lenin °* TB/1375
J. H. HEXTER: More's Utopia: *The Biography of an Idea. Epilogue by the Author* TB/1195
H. STUART HUGHES: History as Art and as Science: *Twin Vistas on the Past* TB/1207
ARTHUR O. LOVEJOY: The Great Chain of Being: *A Study of the History of an Idea* TB/1009
JOSE ORTEGA Y GASSET: The Modern Theme. *Introduction by Jose Ferrater Mora* TB/1038

RICHARD H. POPKIN: The History of Scepticism from Erasmus to Descartes. *Revised Edition* TB/139
G. J. RENIER: History: *Its Purpose and Method* TB/1209
MASSIMO SALVADORI, Ed.: Modern Socialism # HR/137
GEORG SIMMEL , et al.: Essays on Sociology Philosophy and Aesthetics. *Edited by Kur H. Wolff* TB/1234
BRUNO SNELL: The Discovery of the Mind: *The Greek Origins of European Thought* TB/1018
W. WARREN WAGER, ed.: European Intellectua History Since Darwin and Marx TB/129
W. H. WALSH: Philosophy of History: In Introduction TB/1020

History: General

HANS KOHN: The Age of Nationalism: *The First Era of Global History* TB/1380
BERNARD LEWIS: The Arabs in History TB/1029
BERNARD LEWIS: The Middle East and the West ° TB/1274

History: Ancient

A. ANDREWS: The Greek Tyrants TB/1103
ERNST LUDWIG EHRLICH: A Concise History o Israel: *From the Earliest Times to the Destruction of the Temple in A.D. 70 °* TB/128
ADOLF ERMAN, Ed.: The Ancient Egyptians: *A Sourcebook of their Writings. New Introduction by William Kelly Simpson* TB/123
THEODOR H. GASTER: Thespis: *Ritual Myth an Drama in the Ancient Near East* TB/128
MICHAEL GRANT: Ancient History ° TB/1190
A. H. M. JONES, Ed.: A History of Rom through the Fifgth Century # *Vol. I: The Republic* HR/136
Vol. II The Empire: HR/146
SAMUEL NOAH KRAMER: Sumerian Mythology TB/105
NAPHTALI LEWIS & MEYER REINHOLD, Eds. Roman Civilization *Vol. I: The Republic* TB/123
Vol. II: The Empire TB/123

History: Medieval

MARSHALL W. BALDWIN, Ed.: Christianit Through the 13th Century # HR/146
MARC BLOCH: Land and Work in Medieva Europe. *Translated by J. E. Anderson* TB/1452
HELEN CAM: England Before Elizabeth TB/1026
NORMAN COHN: The Pursuit of the Millennium Revolutionary Messianism in Medieval an Reformation Europe TB/103
G. G. COULTON: Medieval Village, Manor, an Monastery HR/102
HEINRICH FICHTENAU: The Carolingian Empire *The Age of Charlemagne. Translated with a Introduction by Peter Munz* TB/114
GALBERT OF BRUGES: The Murder of Charles th Good: *A Contemporary Record of Revolu tionary Change in 12th Century Flanders Translated with an Introduction by Jame Bruce Ross* TB/131
F. L. GANSHOF: Feudalism TB/105
F. L. GANSHOF: The Middle Ages: *A History o International Relations. Translated by Rém Hall* TB/141
W. O. HASSALL, Ed.: Medieval England: *A Viewed by Contemporaries* TB/120
DENYS HAY: The Medieval Centuries ° TB/119
DAVID HERLIHY, Ed.: Medieval Culture and Sc citey # HR/134

4

M. HUSSEY: The Byzantine World TB/1057
ЭBERT LATOUCHE: The Birth of Western Economy: *Economic Aspects of the Dark Ages* ° TB/1290
ENRY CHARLES LEA: The Inquisition of the Middle Ages. || *Introduction by Walter Ullmann* TB/1456
ЭRDINARD LOT: The End of the Ancient World and the Beginnings of the Middle Ages. *Introduction by Glanville Downey* TB/1044
, R. LOYN: The Norman Conquest TB/1457
CHILLE LUCHAIRE: Social France at the time of Philip Augustus. *Intro. by John W. Baldwin* TB/1314
JIBERT DE NOGENT: Self and Society in Medieval France: *The Memoirs of Guibert de Nogent.* || Edited by John F. Benton TB/1471
'ARSILIUS OF PADUA: The Defender of Peace. *The Defensor Pacis. Translated with an Introduction by Alan Gewirth* TB/1310
HARLES PETET-DUTAILLIS: The Feudal Monarchy in France and England: *From the Tenth to the Thirteenth Century* ° TB/1165
TEVEN RUNCIMAN: A History of the Crusades Vol. I: *The First Crusade and the Foundation of the Kingdom of Jerusalem. Illus.* TB/1143
Vol. II: *The Kingdom of Jerusalem and the Frankish East 1100-1187. Illus.* TB/1243
Vol. III: *The Kingdom of Acre and the Later Crusades. Illus.* TB/1298
M. WALLACE-HADRILL: The Barbarian West: *The Early Middle Ages, A.D. 400-1000* TB/1061

'istory: Renaissance & Reformation

ACOB BURCKHARDT: The Civilization of the Renaissance in Italy. *Introduction by Benjamin Nelson and Charles Trinkaus. Illus.* Vol. I TB/40; Vol. II TB/41
ЭHN CALVIN & JACOPO SADOLETO: A Reformation Debate. *Edited by John C. Olin* TB/1239
EDERICO CHABOD: Machiavelli and the Renaissance TB/1193
ЧOMAS CROMWELL: Thomas Cromwell on Church and Commonwealth,: *Selected Letters 1523-1540.* ¶ *Ed. with an Intro. by Arthur J. Slavin* TB/1462
, TREVOR DAVIES: The Golden Century of Spain, 1501-1621 ° TB/1194
, H. ELLIOTT: Europe Divided, 1559-1598 a ° TB/1414
, R. ELTON: Reformation Europe, 1517-1559 ° a TB/1270
ESIDERIUS ERASMUS: Christian Humanism and the Reformation: *Selected Writings. Edited and Translated by John C. Olin* TB/1166
ESIDERIUS ERASMUS: Erasmus and His Age: *Selected Letters. Edited with an Introduction by Hans J. Hillerbrand. Translated by Marcus A. Haworth* TB/1461
'ALLACE K. FERGUSON et al.: Facets of the Renaissance TB/1098
'ALLACE K. FERGUSON et al.: The Renaissance: *Six Essays. Illus.* TB/1084
RANCESCO GUICCIARDINI: History of Florence. *Translated with an Introduction and Notes by Mario Domandi* TB/1470
'ERNER L. GUNDERSHEIMER, Ed.: French Humanism, 1470-1600. * *Illus.* TB/1473
:ARIE BOAS HALL, Ed.: Nature and Nature's Laws: *Documents of the Scientific Revolution* # HR/1420
ANS J. HILLERBRAND, Ed., The Protestant Reformation # HR/1342
ЭHAN HUIZINGA: Erasmus and the Age of Reformation. *Illus.* TB/19

JOEL HURSTFIELD: The Elizabethan Nation TB/1312
JOEL HURSTFIELD, Ed.: The Reformation Crisis TB/1267
PAUL OSKAR KRISTELLER: Renaissance Thought: *The Classic, Scholastic, and Humanist Strains* TB/1048
PAUL OSKAR KRISTELLER: Renaissance Thought II: *Papers on Humanism and the Arts* TB/1163
PAUL O. KRISTELLER & PHILIP P. WIENER, Eds.: Renaissance Essays TB/1392
DAVID LITTLE: Religion, Order and Law: *A Study in Pre-Revolutionary England.* § *Preface by R. Bellah* TB/1418
NICCOLO MACHIAVELLI: History of Florence and of the Affairs of Italy: *From the Earliest Times to the Death of Lorenzo the Magnificent. Introduction by Felix Gilbert* TB/1027
ALFRED VON MARTIN: Sociology of the Renaissance. ° *Introduction by W. K. Ferguson* TB/1099
GARRETT MATTINGLY et al.: Renaissance Profiles. *Edited by J. H. Plumb* TB/1162
J. E. NEALE: The Age of Catherine de Medici ° TB/1085
J. H. PARRY: The Establishment of the European Hegemony: 1415-1715: *Trade and Exploration in the Age of the Renaissance* TB/1045
J. H. PARRY, Ed.: The European Reconnaissance: *Selected Documents* # HR/1345
BUONACCORSO PITTI & GREGORIO DATI: Two Memoirs of Renaissance Florence: *The Diaries of Buonaccorso Pitti and Gregorio Dati. Edited with Intro. by Gene Brucker. Trans. by Julia Martines* TB/1333
J. H. PLUMB: The Italian Renaissance: *A Concise Survey of Its History and Culture* TB/1161
A. F. POLLARD: Henry VIII. *Introduction by A. G. Dickens.* ° TB/1249
RICHARD H. POPKIN: The History of Scepticism from Erasmus to Descartes TB/139
PAOLO ROSSI: Philosophy, Technology, and the Arts, in the Early Modern Era 1400-1700. || *Edited by Benjamin Nelson. Translated by Salvator Attanasio* TB/1458
FERDINAND SCHEVILL: The Medici. *Illus.* TB/1010
FERDINAND SCHEVILL: Medieval and Renaissance Florence. *Illus.* Vol. I: *Medieval Florence* TB/1090
Vol. II: *The Coming of Humanism and the Age of the Medici* TB/1091
R. H. TAWNEY: The Agrarian Problem in the Sixteenth Century. *Intro. by Lawrence Stone* TB/1315
H. R. TREVOR-ROPER: The European Witch-craze of the Sixteenth and Seventeenth Centuries and Other Essays ° TB/1416
VESPASIANO: Rennaissance Princes, Popes, and XVth Century: *The Vespasiano Memoirs. Introduction by Myron P. Gilmore. Illus.* TB/1111

History: Modern European

RENE ALBRECHT-CARRIE, Ed.: The Concert of Europe # HR/1341
MAX BELOFF: The Age of Absolutism, 1660-1815 TB/1062
OTTO VON BISMARCK: Reflections and Reminiscences. *Ed. with Intro. by Theodore S. Hamerow* ¶ TB/1357
EUGENE C. BLACK, Ed.: British Politics in the Nineteenth Century # HR/1427

EUGENE C. BLACK, Ed.: European Political History, 1815-1870: *Aspects of Liberalism* ¶ TB/1331
ASA BRIGGS: The Making of Modern England, 1783-1867: *The Age of Improvement* ° TB/1203
D. W. BROGAN: The Development of Modern France ° Vol. I: *From the Fall of the Empire to the Dreyfus Affair* TB/1184 Vol. II: *The Shadow of War, World War I, Between the Two Wars* TB/1185
ALAN BULLOCK: Hitler, A Study in Tyranny. ° *Revised Edition. Illus.* TB/1123
EDMUND BURKE: On Revolution. *Ed. by Robert A. Smith* TB/1401
E. R. CARR: International Relations Between the Two World Wars, 1919-1939 ° TB/1279
E. H. CARR: The Twenty Years' Crisis, 1919-1939: *An Introduction to the Study of International Relations* ° TB/1122
GORDON A. CRAIG: From Bismarck to Adenauer: *Aspects of German Statecraft. Revised Edition* TB/1171
LESTER G. CROCKER, Ed.: The Age of Enlightenment # HR/1423
DENIS DIDEROT: The Encyclopedia: *Selections. Edited and Translated with Introduction by Stephen Gendzier* TB/1299
JACQUES DROZ: Europe between Revolutions, 1815-1848. ° *a Trans. by Robert Baldick* TB/1346
JOHANN GOTTLIEB FICHTE: Addresses to the German Nation. *Ed. with Intro. by George A. Kelly* ¶ TB/1366
FRANKLIN L. FORD: Robe and Sword: *The Re-Louis XIV* TB/1217
ROBERT & ELBORG FORSTER, Eds.: European Society in the Eighteenth Century # HR/1404
C. C. GILLISPIE: Genesis and Geology: *The Decades before Darwin* § TB/51
ALBERT GOODWIN, Ed.: The European Nobility in the Enghteenth Century TB/1313
ALBERT GOODWIN: The French Revolution TB/1064
ALBERT GUERARD: France in the Classical Age: *The Life and Death of an Ideal* TB/1183
JOHN B. HALSTED, Ed.: Romanticism # HR/1387
J. H. HEXTER: Reappraisals in History: *New Views on History and Society in Early Modern Europe* ° TB/1100
STANLEY HOFFMANN et al.: In Search of France: *The Economy, Society and Political System In the Twentieth Century* TB/1219
H. STUART HUGHES: The Obstructed Path: *French Social Thought in the Years of Desperation* TB/1451
JOHAN HUIZINGA: Dutch Civilisation in the 17th Century and Other Essays TB/1453
LIONAL KOCHAN: The Struggle for Germany: *1914-45* TB/1304
HANS KOHN: The Mind of Germany: *The Education of a Nation* TB/1204
HANS KOHN, Ed.: The Mind of Modern Russia: *Historical and Political Thought of Russia's Great Age* TB/1065
WALTER LAQUEUR & GEORGE L. MOSSE, Eds.: Education and Social Structure in the 20th Century. ° *Volume 6 of the Journal of Contemporary History* TB/1339
WALTER LAQUEUR & GEORGE L. MOSSE, Ed.: International Fascism, 1920-1945. ° *Volume 1 of the Journal of Contemporary History* TB/1276
WALTER LAQUEUR & GEORGE L. MOSSE, Eds.: Literature and Politics in the 20th Century. ° *Volume 5 of the Journal of Contemporary History.* TB/1328

WALTER LAQUEUR & GEORGE L. MOSSE, Eds.: Th New History: *Trends in Historical Researc and Writing Since World War II.* ° *Volum 4 of the* Journal of Contemporary History TB/132
WALTER LAQUEUR & GEORGE L. MOSSE, Eds. 1914: *The Coming of the First World War* ° *Volume3 of the* Journal of Contemporar History TB/130
C. A. MACARTNEY, Ed.: The Habsburg an Hohenzollern Dynasties in the Seventeenth and Eighteenth Centuries # HR/140
JOHN MCMANNERS: European History, 1789 1914: *Men, Machines and Freedom* TB/141
PAUL MANTOUX: The Industrial Revolution in the Eighteenth Century: *An Outline of th Beginnings of the Modern Factory System in England* TB/107
FRANK E. MANUEL: The Prophets of Paris: *Tur got, Condorcet, Saint-Simon, Fourier, an Comte* TB/121
KINGSLEY MARTIN: French Liberal Thought i the Eighteenth Century: *A Study of Politico Ideas from Bayle to Condorcet* TB/111
NAPOLEON III: Napoleonic Ideas: *Des Idée Napoléoniennes, par le Prince Napoléon-Loui Bonaparte. Ed. by Brison D. Gooch* ¶ TB/133
FRANZ NEUMANN: Behemoth: *The Structure an Practice of National Socialism, 1933-1944* TB/128
DAVID OGG: Europe of the Ancien Régime, 1715 1783 ° *a* TB/127
GEORGE RUDE: Revolutionary Europe, 1783 1815 ° *a* TB/127
MASSIMO SALVADORI, Ed.: Modern Socialism ≠ TB/137
HUGH SETON-WATSON: Eastern Europe Betwee the Wars, 1918-1941 TB/133
DENIS MACK SMITH, Ed.: The Making of Italy 1796-1870 # HR/135
ALBERT SOREL: Europe Under the Old Regime *Translated by Francis H. Herrick* TB/112
ROLAND N. STROMBERG, Ed.: Realism, Natural ism, and Symbolism: *Modes of Thought an Expression in Europe, 1848-1914* # HR/135
A. J. P. TAYLOR: From Napoleon to Lenin: *His torical Essays* ° TB/126
A. J. P. TAYLOR: The Habsburg Monarchy, 1809 1918: *A History of the Austrian Empire an Austria-Hungary* ° TB/118
J. M. THOMPSON: European History, 1494-178 TB/143
DAVID THOMSON, Ed.: France: *Empire and Re public, 1850-1940* # HR/138
ALEXIS DE TOCQUEVILLE & GUSTAVE DE BEAUMONT Tocqueville and Beaumont on Social Reform *Ed. and trans. with Intro. by Seymou Drescher* TB/134
G. M. TREVELYAN: British History in the Nine teenth Century and After: 1792-1919 ° TB/125
H. R. TREVOR-ROPER: Historical Essays TB/126
W. WARREN WAGAR, Ed.: Science, Faith, an MAN: *European Thought Since 1914* # HR/136
MACK WALKER, Ed.: Metternich's Europe, 1813 1848 # HR/136
ELIZABETH WISKEMANN: Europe of the Dictators 1919-1945 ° *a* TB/127
JOHN B. WOLF: France: 1814-1919: *The Rise c a Liberal-Democratic Society* TB/301

Literature & Literary Criticism

JACQUES BARZUN: The House of Intellect TB/105

. J. BATE: From Classic to Romantic: *Premises of Taste in Eighteenth Century England* TB/1036

AN WYCK BROOKS: Van Wyck Brooks: The Early Years: *A Selection from his Works, 1908-1921* Ed. with Intro. by Claire Sprague TB/3082

RNST R. CURTIUS: European Literature and the Latin Middle Ages. *Trans. by Willard Trask* TB/2015

CHMOND LATTIMORE, Translator: The Odyssey of Homer TB/1389

OHN STUART MILL: On Bentham and Coleridge. *Introduction by F. R. Leavis* TB/1070

AMUEL PEPYS: The Diary of Samuel Pepys. ° *Edited by O. F. Morshead. 60 illus. by Ernest Shepard* TB/1007

OBERT PREYER, Ed.: Victorian Literature ** TB/1302

LBION W. TOURGEE: A Fool's Errand: *A Novel of the South during Reconstruction. Intro. by George Fredrickson* TB/3074

ASIL WILEY: Nineteenth Century Studies: *Coleridge to Matthew Arnold* ° TB/1261

AYMOND WILLIAMS: Culture and Society, 1780-1950 ° TB/1252

Philosophy

ENRI BERGSON: Time and Free Will: *An Essay on the Immediate Data of Consciousness* ° TB/1021

UDWIG BINSWANGER: Being-in-the-World: *Selected Papers. Trans. with Intro. by Jacob Needleman* TB/1365

. J. BLACKHAM: Six Existentialist Thinkers: *Kierkegaard, Nietzsche, Jaspers, Marcel, Heidegger, Sartre* ° TB/1002

M. BOCHENSKI: The Methods of Contemporary Thought. *Trans. by Peter Caws* TB/1377

RANE BRINTON: Nietzsche. *Preface, Bibliography, and Epilogue by the Author* TB/1197

RNST CASSIRER: Rousseau, Kant and Goethe. *Intro. by Peter Gay* TB/1092

REDERICK COPLESTON, S. J.: Medieval Philosophy TB/376

M. CORNFORD: From Religion to Philosophy: *A Study in the Origins of Western Speculation* § TB/20

ILFRID DESAN: The Tragic Finale: *An Essay on the Philosophy of Jean-Paul Sartre* TB/1030

MARVIN FARBER: The Aims of Phenomenology: *The Motives, Methods, and Impact of Husserl's Thought* TB/1291

ARVIN FARBER: Basic Issues of Philosophy: *Experience, Reality, and Human Values* TB/1344

ARVIN FARBER: Phenomenology and Existence: *Towards a Philosophy within Nature* TB/1295

AUL FRIEDLANDER: Plato: *An Introduction* TB/2017

ICHAEL GELVEN: A Commentary on Heidegger's "Being and Time" TB/1464

GLENN GRAY: Hegel and Greek Thought TB/1409

. K. C. GUTHRIE: The Greek Philosophers: *From Thales to Aristotle* ° TB/1008

W. F. HEGEL: On Art, Religion Philosophy: *Introductory Lectures to the Realm of Absolute Spirit.* || *Edited with an Introduction by J. Glenn Gray* TB/1463

W. F. HEGEL: Phenomenology of Mind. ° || *Introduction by George Lichtheim* TB/1303

ARTIN HEIDEGGER: Discourse on Thinking. *Translated with a Preface by John M. Anderson and E. Hans Freund. Introduction by John M. Anderson* TB/1459

F. H. HEINEMANN: Existentialism and the Modern Predicament TB/28

WERER HEISENBERG: Physics and Philosophy: *The Revolution in Modern Science. Intro. by F. S. C. Northrop* TB/549

EDMUND HUSSERL: Phenomenology and the Crisis of Philosophy. § *Translated with an Introduction by Quentin Lauer* TB/1170

IMMANUEL KANT: Groundwork of the Metaphysic of Morals. *Translated and Analyzed by H. J. Paton* TB/1159

IMMANUEL KANT: Lectures on Ethics. § *Introduction by Lewis White Beck* TB/105

WALTER KAUFMANN, Ed.: Religion From Tolstoy to Camus: *Basic Writings on Religious Truth and Morals* TB/123

QUENTIN LAUER: Phenomenology: *Its Genesis and Prospect. Preface by Aron Gurwitsch* TB/1169

MAURICE MANDELBAUM: The Problem of Historical Knowledge: *An Answer to Relativism* TB/1338

GEORGE A. MORGAN: What Nietzsche Means TB/1198

H. J. PATON: The Categorical Imperative: *A Study in Kant's Moral Philosophy* TB/1325

MICHAEL POLANYI: Personal Knowledge: *Towards a Post-Critical Philosophy* TB/1158

KARL R. POPPER: Conjectures and Refutations: *The Growth of Scientific Knowledge* TB/1376

WILLARD VAN ORMAN QUINE: Elementary Logic *Revised Edition* TB/577

WILLARD VAN ORMAN QUINE: From a Logical Point of View: *Logico-Philosophical Essays* TB/566

JOHN E. SMITH: Themes in American Philosophy: *Purpose, Experience and Community* TB/1466

MORTON WHITE: Foundations of Historical Knowledge TB/1440

WILHELM WINDELBAND: A History of Philosophy *Vol. I: Greek, Roman, Medieval* TB/38 *Vol. II: Renaissance, Enlightenment, Modern* TB/39

LUDWIG WITTGENSTEIN: The Blue and Brown Books ° TB/1211

LUDWIG WITTGENSTEIN: Notebooks, 1914-1916 TB/1441

Political Science & Government

C. E. BLACK: The Dynamics of Modernization: *A Study in Comparative History* TB/1321

KENNETH E. BOULDING: Conflict and Defense: *A General Theory of Action* TB/3024

DENIS W. BROGAN: Politics in America. *New Introduction by the Author* TB/1469

CRANE BRINTON: English Political Thought in the Nineteenth Century TB/1071

ROBERT CONQUEST: Power and Policy in the USSR: *The Study of Soviet Dynastics* ° TB/1307

ROBERT A. DAHL & CHARLES E. LINDBLOM: Politics, Economics, and Welfare: *Planning and Politico-Economic Systems Resolved into Basic Social Processes* TB/1277

HANS KOHN: Political Ideologies of the 20th Century TB/1277

ROY C. MACRIDIS, Ed.: Political Parties: *Contemporary Trends and Ideas* ** TB/1322

ROBERT GREEN MC CLOSKEY: American Conservatism in the Age of Enterprise, 1865-1910 TB/1137

MARSILIUS OF PADUA: The Defender of Peace. *The Defensor Pacis. Translated with an Introduction by Alan Gewirth* TB/1310

KINGSLEY MARTIN: French Liberal Thought in the Eighteenth Century: *A Study of Political Ideas from Bayle to· Condorcet* TB/1114

BARRINGTON MOORE, JR.: Political Power and Social Theory: *Seven Studies* || TB/1221

BARRINGTON MOORE, JR.: Soviet Politics—The Dilemma of Power: *The Role of Ideas in Social Change* || TB/1222

BARRINGTON MOORE, JR.: Terror and Progress—USSR: *Some Sources of Change and Stability*

JOHN B. MORRALL: Political Thought in Medieval Times TB/1076

KARL R. POPPER: The Open Society and Its Enemies *Vol. I: The Spell of Plato* TB/1101 *Vol. II: The High Tide of Prophecy: Hegel, Marx, and the Aftermath* TB/1102

CONYERS READ, Ed.: The Constitution Reconsidered. *Revised Edition, Preface by Richard B. Morris* TB/1384

JOHN P. ROCHE, Ed.: Origins of American Political Thought: *Selected Readings* TB/1301

JOHN P. ROCHE, Ed.: American Political Thought: *From Jefferson to Progressivism* TB/1332

HENRI DE SAINT-SIMON: Social Organization, The Science of Man, and Other Writings. || *Edited and Translated with an Introduction by Felix Markham* TB/1152

CHARLES SCHOTTLAND, Ed.: The Welfare State ** TB/1323

JOSEPH A. SCHUMPETER: Capitalism, Socialism and Democracy TB/3008

PETER WOLL, Ed.: Public Administration and Policy: *Selected Essays* TB/1284

Psychology

ALFRED ADLER: The Individual Psychology of Alfred Adler: *A Systematic Presentation in Selections from His Writings. Edited by Heinz L. & Rowena R. Ansbacher* TB/1154

ALFRED ADLER: Problems of Neurosis: *A Book of Case Histories. Introduction by Heinz L. Ansbacher* TB/1145

LUDWIG BINSWANGER: Being-in-the-World: *Selected Papers. || Trans. with Intro. by Jacob Needleman* TB/1365

ARTHUR BURTON & ROBERT E. HARRIS: Clinical Studies of Personality Vol. I TB/3075 Vol. II TB/3076

HADLEY CANTRIL: The Invasion from Mars: *A Study in the Psychology of Panic* || TB/1282

MIRCEA ELIADE: Cosmos and History: *The Myth of the Eternal Return* § TB/2050

MIRCEA ELIADE: Myth and Reality TB/1369

MIRCEA ELIADE: Myths, Dreams and Mysteries: *The Encounter Between Contemporary Faiths and Archaic Realities* § TB/1320

MIRCEA ELIADE: Rites and Symbols of Initiation: *The Mysteries of Birth and Rebirth* § TB/1236

HERBERT FINGARETTE: The Self in Transformation: *Psychoanalysis, Philosophy and the Life of the Spirit* || TB/1177

SIGMUND FREUD: On Creativity and the Unconscious: *Papers on the Psychology of Art, Literature, Love, Religion. § Intro. by Benjamin Nelson* TB/45

J. GLENN GRAY: The Warriors: *Reflections on Men in Battle. Introduction by Hannah Arendt* TB/1294

WILLIAM JAMES: Psychology: *The Briefer Course. Edited with an Intro. by Gordon Allport* TB/1034

C. G. JUNG: Psychological Reflections. *Ed. by J. Jacobi* TB/2001

KARL MENNINGER, M.D.: Theory of Psychoanalytic Technique TB/1144

JOHN H. SCHAAR: Escape from Authority: *The Perspectives of Erich Fromm* TB/1155

MUZAFER SHERIF: The Psychology of Social Norms. *Introduction by Gardner Murphy* TB/307

HELLMUT WILHELM: Change: *Eight Lectures on the I Ching* TB/201

Religion: Ancient and Classical, Biblical and Judaic Traditions

W. F. ALBRIGHT: The Biblical Period from Abraham to Ezra TB/10

SALO W. BARON: Modern Nationalism and Religion TB/81

C. K. BARRETT, Ed.: The New Testament Background: *Selected Documents* TB/8

MARTIN BUBER: Eclipse of God: *Studies in the Relation Between Religion and Philosophy* TB/1

MARTIN BUBER: Hasidism and Modern Man *Edited and Translated by Maurice Friedman* TB/83

MARTIN BUBER: The Knowledge of Man. *Edited with an Introduction by Maurice Friedman Translated by Maurice Friedman and Ronald Gregor Smith* TB/13

MARTIN BUBER: Moses. *The Revelation and the Covenant* TB/83

MARTIN BUBER: The Origin and Meaning of Hasidism. *Edited and Translated by Maurice Friedman* TB/83

MARTIN BUBER: The Prophetic Faith TB/7

MARTIN BUBER: Two Types of Faith: *Interpenetration of Judaism and Christianity* ° TB/7

MALCOLM L. DIAMOND: Martin Buber: *Jewish Existentialist* TB/84

M. S. ENSLIN: Christian Beginnings TB/

M. S. ENSLIN: The Literature of the Christian Movement TB/

ERNST LUDWIG EHRLICH: A Concise History of Israel: *From the Earliest Times to the Destruction of the Temple in A.D. 70* ° TB/12

HENRI FRANKFORT: Ancient Egyptian Religion *An Interpretation* TB/7

MAURICE S. FRIEDMAN: Martin Buber: *The Life of Dialogue* TB/6

ABRAHAM HESCHEL: The Earth Is the Lord's & The Sabbath. *Two Essays* TB/82

ABRAHAM HESCHEL: God in Search of Man: *A Philosophy of Judaism* TB/80

ABRAHAM HESCHEL: Man Is not Alone: *A Philosophy of Religion* TB/83

ABRAHAM HESCHEL: The Prophets: *An Introduction* TB/142

T. J. MEEK: Hebrew Origins TB/6

JAMES MUILENBURG: The Way of Israel: *Biblical Faith and Ethics* TB/13

H. J. ROSE: Religion in Greece and Rome TB/5

H. H. ROWLEY: The Growth of the Old Testament TB/10

D. WINTON THOMAS, Ed.: Documents from Old Testament Times TB/8

Religion: General Christianity

ROLAND H. BAINTON: Christendom: *A Short History of Christianity and Its Impact on Western Civilization. Illus.*
Vol. I TB/131; Vol. II TB/132

JOHN T. MCNEILL: Modern Christian Movements. *Revised Edition* TB/140

ERNST TROELTSCH: The Social Teaching of the Christian Churches. *Intro. by H. Richard Niebuhr* Vol. I TB/71; Vol. II TB/72

8

Religion: Early Christianity Through Reformation

ANSELM OF CANTERBURY: Truth, Freedom, and Evil: *Three Philosophical Dialogues. Edited and Translated by Jasper Hopkins and Herbert Richardson* TB/317
MARSHALL W. BALDWIN, Ed.: Christianity through the 13th Century # HR/1468
W. D. DAVIES: Paul and Rabbinic Judaism: *Some Rabbinic Elements in Pauline Theology. Revised Edition* ° TB/146
ADOLF DEISSMAN: Paul: *A Study in Social and Religious History* TB/15
JOHANNES ECKHART: Meister Eckhart: *A Modern Translation by R. Blakney* TB/8
EDGAR J. GOODSPEED: A Life of Jesus TB/1
ROBERT M. GRANT: Gnosticism and Early Christianity TB/136
WILLIAM HALLER: The Rise of Puritanism TB/22
GERHART B. LADNER: The Idea of Reform: *Its Impact on the Christian Thought and Action in the Age of the Fathers* TB/149
ARTHUR DARBY NOCK: Early Gentile Christianity and Its Hellenistic Background TB/111
ARTHUR DARBY NOCK: St. Paul ° TR/104
ORIGEN: On First Principles. *Edited by G. W. Butterworth. Introduction by Henri de Lubac* TB/311
GORDON RUPP: Luther's Progress to the Diet of Worms ° TB/120

Religion: The Protestant Tradition

KARL BARTH: Church Dogmatics: *A Selection. Intro. by H. Gollwitzer. Ed. by G. W. Bromiley* TB/95
KARL BARTH: Dogmatics in Outline TB/56
KARL BARTH: The Word of God and the Word of Man TB/13
HERBERT BRAUN, et al.: God and Christ: *Existence and Province. Volume 5 of Journal for Theology and the Church, edited by Robert W. Funk and Gerhard Ebeling* TB/255
WHITNEY R. CROSS: The Burned-Over District: *The Social and Intellectual History of Enthusiastic Religion in Western New York, 1800-1850* TB/1242
NELS F. S. FERRE: Swedish Contributions to Modern Theology. *New Chapter by William A. Johnson* TB/147
WILLIAM R. HUTCHISON, Ed.: American Protestant Thought: *The Liberal Era* ‡ TB/1385
ERNST KASEMANN, et al.: Distinctive Protestant and Catholic Themes Reconsidered. *Volume 3 of Journal for Theology and the Church, edited by Robert W. Funk and Gerhard Ebeling* TB/253
SOREN KIERKEGAARD: On Authority and Revelation: *The Book on Adler, or a Cycle of Ethico-Religious Essays. Introduction by F. Sontag* TB/139
SOREN KIERKEGAARD: Crisis in the Life of an Actress, *and Other Essays on Drama. Translated with an Introduction by Stephen Crites* TB/145
SOREN KIERKEGAARD: Edifying Discourses. *Edited with an Intro. by Paul Holmer* TB/32
SOREN KIERKEGAARD: The Journals of Kierkegaard. ° *Edited with an Intro. by Alexander Dru* TB/52
SOREN KIERKEGAARD: The Point of View for My Work as an Author: *A Report to History.* § *Preface by Benjamin Nelson* TB/88

SOREN KIERKEGAARD: The Present Age. § *Translated and edited by Alexander Dru. Introduction by Walter Kaufmann* TB/94
SOREN KIERKEGAARD: Purity of Heart. *Trans. by Douglas Steere* TB/4
SOREN KIERKEGAARD: Repetition: *An Essay in Experimental Psychology* § TB/117
SOREN KIERKEGAARD: Works of Love: *Some Christian Reflections in the Form of Discourses* TB/122
WILLIAM G. MCLOUGHLIN, Ed.: The American Evangelicals: 1800-1900: *An Anthology* TB/1382
WOLFHART PANNENBERG, et al.: History and Hermeneutic. *Volume 4 of Journal for Theology and the Church, edited by Robert W. Funk and Gerhard Ebeling* TB/254
JAMES M. ROBINSON, et al.: The Bultmann School of Biblical Interpretation: New Directions? *Volume 1 of Journal for Theology and the Church, edited by Robert W. Funk and Gerhard Ebeling* TB/251
F. SCHLEIERMACHER: The Christian Faith. *Introduction by Richard R. Niebuhr.* Vol. I TB/108; Vol. II TB/109
F. SCHLEIERMACHER: On Religion: *Speeches to Its Cultured Despisers. Intro. by Rudolf Otto* TB/36
TIMOTHY L. SMITH: Revivalism and Social Reform: *American Protestantism on the Eve of the Civil War* TB/1229
PAUL TILLICH: Dynamics of Faith TB/42
PAUL TILLICH: Morality and Beyond TB/142
EVELYN UNDERHILL: Worship TB/10

Religion: The Roman & Eastern Christian Traditions

A. ROBERT CAPONIGRI, Ed.: Modern Catholic Thinkers II: *The Church and the Political Order* TB/307
G. P. FEDOTOV: The Russian Religious Mind: *Kievan Christianity, the tenth to the thirteenth Centuries* TB/370
GABRIEL MARCEL: Being and Having: *An Existential Diary. Introduction by James Collins* TB/310
GABRIEL MARCEL: Homo Viator: *Introduction to a Metaphysic of Hope* TB/397

Religion: Oriental Religions

TOR ANDRAE: Mohammed: *The Man and His Faith* § TB/62
EDWARD CONZE: Buddhism: *Its Essence and Development.* ° *Foreword by Arthur Waley* TB/58
EDWARD CONZE: Buddhist Meditation TB/1442
EDWARD CONZE et al, Editors: Buddhist Texts through the Ages TB/113
ANANDA COOMARASWAMY: Buddha and the Gospel of Buddhism TB/119
H. G. CREEL: Confucius and the Chinese Way TB/63
FRANKLIN EDGERTON, Trans. & Ed.: The Bhagavad Gita TB/115
SWAMI NIKHILANANDA, Trans. & Ed.: The Upanishads TB/114
D. T. SUZUKI: On Indian Mahayana Buddhism. ° *Ed. with Intro. by Edward Conze.* TB/1403

Religion: Philosophy, Culture, and Society

NICOLAS BERDYAEV: The Destiny of Man TB/61
RUDOLF BULTMANN: History and Eschatology: *The Presence of Eternity* ° TB/91
RUDOLF BULTMANN AND FIVE CRITICS: Kerygma and Myth: *A Theological Debate* TB/80

9

RUDOLF BULTMANN and KARL KUNDSIN: Form Criticism: *Two Essays on New Testament Research. Trans. by F. C. Grant* TB/96
WILLIAM A. CLEBSCH & CHARLES R. JAEKLE: Pastoral Care in Historical Perspective: *An Essay with Exhibits* TB/148
FREDERICK FERRE: Language, Logic and God. *New Preface by the Author* TB/1407
LUDWIG FEUERBACH: The Essence of Christianity. *§ Introduction by Karl Barth. Foreword by H. Richard Niebuhr* TB/11
C. C. GILLISPIE: Genesis and Geology: *The Decades before Darwin §* TB/51
ADOLF HARNACK: What Is Christianity? *§ Introduction by Rudolf Bultmann* TB/17
KYLE HASELDEN: The Racial Problem in Christian Perspective TB/116
MARTIN HEIDEGGER: Discourse on Thinking. *Translated with a Preface by John M. Anderson and E. Hans Freund. Introduction by John M. Anderson* TB/1459
IMMANUEL KANT: Religion Within the Limits of Reason Alone. *§ Introduction by Theodore M. Greene and John Silber* TB/FG
WALTER KAUFMANN, Ed.: Religion from Tolstoy to Camus: *Basic Writings on Religious Truth and Morals. Enlarged Edition* TB/123
JOHN MACQUARRIE: An Existentialist Theology: *A Comparison of Heidegger and Bultmann. ° Foreword by Rudolf Bultmann* TB/125
H. RICHARD NIERUHR: Christ and Culture TB/3
H. RICHARD NIEBUHR: The Kingdom of God in America TB/49
ANDERS NYGREN: Agape and Eros. *Translated by Philip S. Watson °* TB/1430
JOHN H. RANDALL, JR.: The Meaning of Religion for Man. *Revised with New Intro. by the Author* TB/1379
WALTER RAUSCHENBUSCHS Christianity and the Social Crisis. *‡ Edited by Robert D. Cross* TB/3059
JOACHIM WACH: Understanding and Believing. *Ed. with Intro. by Joseph M. Kitagawa* TB/1399

Science and Mathematics

JOHN TYLER BONNER: The Ideas of Biology. *Σ Illus.* TB/570
W. E. LE GROS CLARK: The Antecedents of Man: *An Introduction to the Evolution of the Primates. ° Illus.* TB/559
ROBERT E. COKER: Streams, Lakes, Ponds. *Illus.* TB/586
ROBERT E. COKER: This Great and Wide Sea: *An Introduction to Oceanography and Marine Biology. Illus.* TB/551
W. H. DOWDESWELL: Animal Ecology. *61 illus.* TB/543
C. V. DURELL: Readable Relativity. *Foreword by Freeman J. Dyson* TB/530
GEORGE GAMOW: Biography of Physics. *Σ Illus.* TB/567
F. K. HARE: The Restless Atmosphere TB/560
S. KORNER: The Philosophy of Mathematics: *An Introduction* TB/547
J. R. PIERCE: Symbols, Signals and Noise: *The Nature and Process of Communication Σ* TB/574
WILLARD VAN ORMAN QUINE: Mathematical Logic TB/558

Science: History

MARIE BOAS: The Scientific Renaissance, 1450-1630 ° TB/583
W. DAMPIER, Ed.: Readings in the Literature of Science. *Illus.* TB/512

STEPHEN TOULMIN & JUNE GOODFIELD: The Architecture of Matter: *The Physics, Chemistry and Physiology of Matter, Both Animate and Inanimate, as it has Evolved since the Beginnings of Science* TB/584
STEPHEN TOULMIN & JUNE GOODFIELD: The Discovery of Time TB/585
STEPHEN TOULMIN & JUNE GOODFIELD: The Fabric of the Heavens: *The Development of Astronomy and Dynamics* TB/579

Science: Philosophy

J. M. BOCHENSKI: The Methods of Contemporary Thought. *Tr. by Peter Caws* TB/1377
J. BRONOWSKI: Science and Human Values. *Revised and Enlarged. Illus.* TB/505
WERNER HEISENBERG: Physics and Philosophy: *The Revolution in Modern Science. Introduction by F. S. C. Northrop* TB/549
KARL R. POPPER: Conjectures and Refutations: *The Growth of Scientific Knowledge* TB/1376
KARL R. POPPER: The Logic of Scientific Discovery TB/576
STEPHEN TOULMIN: Foresight and Understanding: *An Enquiry into the Aims of Science. Foreword by Jacques Barzun* TB/564
STEPHEN TOULMIN: The Philosophy of Science: *An Introduction* TB/513

Sociology and Anthropology

REINHARD BENDIX: Work and Authority in Industry: *Ideologies of Management in the Course of Industrialization* TB/3035
BERNARD BERELSON, Ed., The Behavioral Sciences Today TB/1127
JOSEPH B. CASAGRANDE, Ed.: In the Company of Man: *Twenty Portraits of Anthropological Informants. Illus.* TB/3047
KENNETH B. CLARK: Dark Ghetto: *Dilemmas of Social Power. Foreword by Gunnar Myrdal* TB/1317
KENNETH CLARK & JEANNETTE HOPKINS: A Relevant War Against Poverty: *A Study of Community Action Programs and Observable Social Change* TB/1480
W. E. LE GROS CLARK: The Antecedents of Man: *An Introduction to the Evolution of the Primates. ° Illus.* TB/559
LEWIS COSER, Ed.: Political Sociology TB/1293
ROSE L. COSER, Ed.: Life Cycle and Achievement in America ** TB/1434
ALLISON DAVIS & JOHN DOLLARD: Children of Bondage: *The Personality Development of Negro Youth in the Urban South ||* TB/3049
ST. CLAIR DRAKE & HORACE R. CAYTON: Black Metropolis: *A Study of Negro Life in a Northern City. Introduction by Everett C. Hughes. Tables, maps, charts, and graphs* Vol. I TB/1086; Vol. II TB/1087
PETER E. DRUCKER: The New Society: *The Anatomy of Industrial Order* TB/1082
CORA DU BOIS: The People of Alor. *With a Preface by the Author* Vol. I *Illus.* TB/1042; Vol. II TB/1043
EMILE DURKHEIM et al.: Essays on Sociology and Philosophy: *with Appraisals of Durkheim's Life and Thought. || Edited by Kurt H. Wolff* TB/1151
LEON FESTINGER, HENRY W. RIECKEN, STANLEY SCHACHTER: When Prophecy Fails: *A Social and Psychological Study of a Modern Group that Predicted the Destruction of the World ||* TB/1132

CHARLES Y. GLOCK & RODNEY STARK: Christian Beliefs and Anti-Semitism. *Introduction by the Authors* TB/1454

ALVIN W. GOULDNER: The Hellenic World TB/1479

ALVIN W. GOULDNER: Wildcat Strike: *A Study in Worker-Management Relationships* || TB/1176

CESAR GRANA: Modernity and Its Discontents: *French Society and the French Man of Letters in the Nineteenth Century* TB/1318

L. S. B. LEAKEY: Adam's Ancestors: *The Evolution of Man and His Culture. Illus.* TB/1019

KURT LEWIN: Field Theory in Social Science. *Selected Theoretical Papers.* || *Edited by Dorwin Cartwright* TB/1135

RITCHIE P. LOWRY: Who's Running This Town? *Community Leadership and Social Change* TB/1383

R. M. MACIVER: Social Causation TB/1153

GARY T. MARX: Protest and Prejudice: *A Study of Belief in the Black Community* TB/1435

ROBERT K. MERTON, LEONARD BROOM, LEONARD S. COTTRELL, JR., Editors: Sociology Today: *Problems and Prospects* ||
Vol. I TB/1173; Vol. II TB/1174

GILBERT OSOFSKY, Ed.: The Burden of Race: A Documentary History of Negro-White Relations in America TB/1405

GILBERT OSOFSKY: Harlem: The Making of a Ghetto: *Negro New York 1890-1930* TB/1381

TALCOTT PARSONS & EDWARD A. SHILS, Editors: Toward a General Theory of Action: *Theoretical Foundations for the Social Sciences* TB/1083

PHILIP RIEFF: The Triumph of the Therapeutic: *Uses of Faith After Freud* TB/1360

JOHN H. ROHRER & MUNRO S. EDMONSON, Eds.: The Eighth Generation Grows Up: *Cultures and Personalities of New Orleans Negroes* || TB/3050

ARNOLD ROSE: The Negro in America: *The Condensed Version of Gunnar Myrdal's* An American Dilemma. *Second Edition* TB/3048

GEORGE ROSEN: Madness in Society: *Chapters in the Historical Sociology of Mental Illness.* || *Preface by Benjamin Nelson* TB/1337

PHILIP SELZNICK: TVA and the Grass Roots: *A Study in the Sociology of Formal Organization* TB/1230

PITIRIM A. SOROKIN: Contemporary Sociological Theories: *Through the First Quarter of the Twentieth Century* TB/3046

MAURICE R. STEIN: The Eclipse of Community: *An Interpretation of American Studies* TB/1128

WILLIAM I. THOMAS: The Unadjusted Girl: *With Cases and Standpoint for Behavior Analysis. Intro. by Michael Parenti* TB/1319

EDWARD A. TIRYAKIAN, Ed.: Sociological Theory, Values and Sociocultural Change: *Essays in Honor of Pitirim A. Sorokin* ° TB/1316

FERDINAND TONNIES: Community and Society: *Gemeinschaft und Gesellschaft. Translated and Edited by Charles P. Loomis* TB/1116

SAMUEL E. WALLACE: Skid Row as a Way of Life TB/1367

W. LLOYD WARNER and Associates: Democracy in Jonesville: *A Study in Quality and Inequality* || TB/1129

W. LLOYD WARNER: Social Class in America: *The Evaluation of Status* TB/1013

FLORIAN ZNANIECKI: The Social Role of the Man of Knowledge. *Introduction by Lewis A. Coser* TB/1372